Introduction to the Finite-Difference Time-Domain (FDTD) Method for Electromagnetics

Synthesis Lectures on Computational Electromagnetics

Editor
Constantine A. Balanis, *Arizona State University*

Synthesis Lectures on Computational Electromagnetics will publish 50- to 100-page publications on topics that include advanced and state-of-the-art methods for modeling complex and practical electromagnetic boundary value problems. Each lecture develops, in a unified manner, the method based on Maxwell's equations along with the boundary conditions and other auxiliary relations, extends underlying concepts needed for sequential material, and progresses to more advanced techniques and modeling. Computer software, when appropriate and available, is included for computation, visualization and design. The authors selected to write the lectures are leading experts on the subject that have extensive background in the theory, numerical techniques, modeling, computations and software development.
The series is designed to:

- Develop computational methods to solve complex and practical electromagnetic boundary-value problems of the 21st century.

- Meet the demands of a new era in information delivery for engineers, scientists, technologists and engineering managers in the fields of wireless communication, radiation, propagation, communication, navigation, radar, RF systems, remote sensing, and biotechnology who require a better understanding and application of the analytical, numerical and computational methods for electromagnetics.

Introduction to the Finite-Difference Time-Domain (FDTD) Method for Electromagnetics
Stephen D. Gedney
2010

Analysis and Design of Substrate Integrated Waveguide Using Efficient 2D Hybrid Method
Xuan Hui Wu, Ahmed A. Kishk
2010

An Introduction to the Locally-Corrected Nyström Method
Andrew F. Peterson, Malcolm M. Bibby
2009

Introduction to the Finite-Difference Time-Domain (FDTD) Method for Electromagnetics

Stephen D. Gedney

ISBN: 978-3-031-00584-8   paperback
ISBN: 978-3-031-01712-4   ebook

DOI 10.1007/978-3-031-01712-4

A Publication in the Springer series
SYNTHESIS LECTURES ON COMPUTATIONAL ELECTROMAGNETICS

Lecture #27
Series Editor: Constantine A. Balanis, *Arizona State University*
Series ISSN
Synthesis Lectures on Computational Electromagnetics
Print 1932-1252   Electronic 1932-1716

Introduction to the Finite-Difference Time-Domain (FDTD) Method for Electromagnetics

Stephen D. Gedney
University of Kentucky

SYNTHESIS LECTURES ON COMPUTATIONAL ELECTROMAGNETICS #27

ABSTRACT

Introduction to the Finite-Difference Time-Domain (FDTD) Method for Electromagnetics provides a comprehensive tutorial of the most widely used method for solving Maxwell's equations – the Finite Difference Time-Domain Method. This book is an essential guide for students, researchers, and professional engineers who want to gain a fundamental knowledge of the FDTD method. It can accompany an undergraduate or entry-level graduate course or be used for self-study. The book provides all the background required to either research or apply the FDTD method for the solution of Maxwell's equations to practical problems in engineering and science. *Introduction to the Finite-Difference Time-Domain (FDTD) Method for Electromagnetics* guides the reader through the foundational theory of the FDTD method starting with the one-dimensional transmission-line problem and then progressing to the solution of Maxwell's equations in three dimensions. It also provides step by step guides to modeling physical sources, lumped-circuit components, absorbing boundary conditions, perfectly matched layer absorbers, and sub-cell structures. Post processing methods such as network parameter extraction and far-field transformations are also detailed. Efficient implementations of the FDTD method in a high level language are also provided.

KEYWORDS

FDTD, Finite-difference time-domain, FDTD software, Electromagnetics, Electrodynamics, Computational Electromagnetics, Maxwell Equations, Transmission line equations, Perfectly Matched Layer, PML, Antennas, RF and microwave engineering

To my wife, Carolyn,
and children Albert, Teresa, Stephanie, David, Joseph, and Mary

Contents

CHAPTER 1

Introduction

1.1 A BRIEF HISTORY OF THE FDTD METHOD

The Finite-Difference Time-Domain (FDTD) method was originally proposed by Kane S. Yee in the seminal paper he published in 1966 [1]. Yee proposed a discrete solution to Maxwell's equations based on central difference approximations of the spatial and temporal derivatives of the curl-equations. The novelty of Yee's approach was the staggering of the electric and magnetic fields in both space and time in order to obtain second-order accuracy. Yee derived a full three-dimensional formulation, and he validated the method with two-dimensional problems. Yee's method went mostly un-noticed for nearly a decade. Finally, in 1975, Taflove and Brodwin applied Yee's method to simulate the scattering by dielectric cylinders [2] and biological heating [3], and in 1977, Holland applied it to predict the currents induced on an aircraft by an electromagnetic pulse (EMP) [4]. Holland summarized the reason for the delay of the application of Yee's method [4]:

> The algorithm ... was first described by Yee about ten years ago. At that time, computers did not exist which could implement this scheme practically in three dimensions. This is no longer the case. It is now possible to mesh a problem space of interest into a $30 \times 30 \times 30$ grid and have random access to the resulting 162,000 field quantities. These advances in computer technology have prompted Longmire to credit Yee's algorithm as being the "fastest numerical way of solving Maxwell's equations [5]."

The growth of the FDTD method since the late 1970's can certainly be paralleled to the ongoing advances in computer technology. In fact, with today's massively parallel computers, a grid dimension of $3,000 \times 3,000 \times 3,000$, which requires nearly 1.5 terabytes of memory, is feasible. Longmire's quote above summarizes another reason for its growth. The FDTD method is a fast algorithm. If N is the total number of degree's of freedom in the three-dimensional space, each time-iteration only requires $O(N)$ floating-point operations. The caveat is that the discrete mesh has to fill the full three-dimensional space. Thus, the number of degrees of freedom grows cubically with the linear dimension of the problem domain.

Another reason for the wide-spread use of the FDTD method is its simplicity. One is reminded of Chef Auguste Gusteau's signature statement in Pixar's movie *Ratatouille*: "Anyone can cook!". Similarly, one can say: "Anyone can FDTD!". The FDTD method is a wonderfully simple method that can be taught at the undergraduate or early graduate level. Yet, it is capable of solving extremely sophisticated engineering problems.

The maturation of the FDTD method can be attributed to a number of pioneers who have made a host of contributions to advance Yee's base FDTD algorithm. A partial list of innovators and their innovations is summarized in Table 1.1. Each innovation has played a key role in advancing the FDTD method to where it is today.

One set of key contributions revolve around the ability of the FDTD method to simulate unbounded problems. To model such problems, the discrete domain must be truncated via a reflectionless absorbing boundary. In 1981, G. Mur proposed a second-order accurate absorbing boundary condition (ABC) [6] that helped to resolve this issue. Later, Higdon introduced a class of boundary operators that was more versatile [7, 8]. Betz-and Mittra extended Higdon's ABC to absorb evanescent waves [9]. While such ABC's were widely used, the dynamic range of the FDTD method as well as the range of applications was still limited by this class of absorbing boundaries. A much more accurate absorbing boundary based on a Perfectly Matched Layer (PML) absorbing medium was proposed by J.-P. Berenger [10, 11]. In fact, the reflection error of the PML can be 3 or 4 orders of magnitude smaller than that offered by an ABC. The PML was also capable of truncating unbounded media that was lossy, inhomogeneous, dispersive, non-linear, or anisotropic. Further developments of Berenger's PML also rendered it capable of absorbing evanescent waves and near fields. The tradeoff of using the PML is an extension of the mesh and additional degrees of freedom in the PML region. Thus, it is more computationally intensive than a local ABC. However, in the mid 1990's, these resources were available even with commodity computers. The direct result of this innovation was that the FDTD method with PML absorbing boundaries was capable of being applied to a much broader range of problems.

One of the advantages of the FDTD solution of Maxwell's equations is that arbitrary media types can inherently be modeled. The FDTD method naturally accommodates inhomogeneous and lossy media. More complex media types, such as frequency-dependent dispersive, anisotropic, bi-anisotropic, chiral, or non-linear media can also be accommodated within the FDTD method. Significant effort has been made by researchers to develop accurate and efficient algorithms to model such media within the context of the FDTD method. A number of seminal works on this topic are listed in Table 1.1. An excellent summary of handling complex media within the FDTD method is provided by F. L. Teixeira in [12].

The basic Yee-algorithm is restricted to a regularly-spaced orthogonal grid. This is not amenable to high-fidelity modeling of very complex geometries. One way this has been mitigated is via the use of subcell modeling techniques. Subcell models make use of local approximations to more accurately resolve the fields near geometric features that are smaller than the local cell dimensions. The first subcell model was introduced as early as 1981 by Holland and Simpson for the modeling of thin wires situated within the FDTD grid [13]. Since then, numerous subcell models were introduced for a variety of applications. Several are listed in Table 1.1. A successful subcell model improves the accuracy of the local fields without sacrificing stability of the algorithm or significantly reducing the time-step. An alternate method of resolving fine geometric features is to introduce local sub-grids. That is, a sub-grid is embedded within the global grid in order to locally resolve fine geometric structure without sacrificing the global space/time scale. The application of subcell models and sub-gridding methods has been quite critical in enhancing the efficiency and the accuracy of the FDTD method for modeling very complex systems.

As an example, consider the model of the cell phone geometry in Fig. 1.1 (from [14]). The

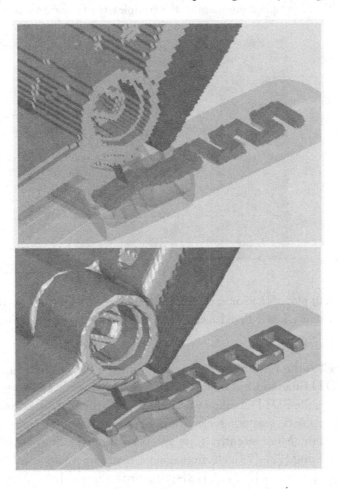

Figure 1.1: FDTD mesh of a flip-phone, illustrating the mesh near the joint and the antenna. Staircased mesh (above), and conformal mesh (below). *Source:* Benkler, et al., *IEEE Trans. on Antennas and Prop.*, 2006, pp. 1843–1849. © 2006 IEEE.

geometry in the upper figure illustrates a staircased orthogonal grid mesh. The mesh in the lower figure was created using a coarser grid with conformal meshing (c.f., Sections 7.3 – 7.4). The conformal mesh was created with SEMCAD X [15]. The antenna was excited with a 1.85 GHz source. Figure 1.2 illustrates the magnitude of the electric field (in dB) in a plane 5 mm below the antenna computed via SEMCAD X using the refined staircased grid (above) and the conformal grid (below) [14]. The agreement is quite good. However, the mesh size of the staircased grid had roughly 6 times more unknowns. The input impedance of the antenna at 1.85 GHz was also computed: $45.8 + j1.07\Omega$, using the refined staircase grid, and $45.6 - j1.45\Omega$, using the coarser conformal grid. This example demonstrates the complexity of the problem that the FDTD method

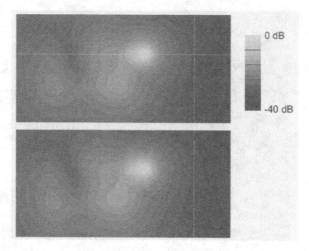

Figure 1.2: Contour plot of $|\vec{E}|$ 5 mm below the antenna at 1.85 GHz. Refined staircase mesh (above), conformal mesh (below). *Source:* Benkler, et al., *IEEE Trans. on Antennas and Prop.*, 2006, pp. 1843-1849. © 2006 IEEE.

can handle, as well as the advantage of using a conformal-FDTD algorithm.

As the FDTD method has matured, the number of application areas it has impacted has diversified. Initially, the FDTD method was applied primarily to classical areas in electromagnetics, including electromagnetic scattering, electromagnetic compatibility, antennas, microwave circuits, and wave propagation. More recently, it has also been applied a diversity of other areas, including biomedical engineering [72]–[75], electromagnetic environmental hazards [76], ground penetrating radar [51, 77], photonics [24, 71], [78]–[80], biophotonics [81, 82], plasmonics [83], photovoltaics [66, 67], nano-optical storage devices [69], and seismic detection [84].

The FDTD method is continuing to develop and expand. Current thrust areas include high-order FDTD methods, more general gridding techniques, unconditionally stable schemes, and multiphysics applications. With these endeavors and many others in progress, as well as continued advances in computing, the FDTD time-line in Table 1.1 will undoubtedly continue to grow.

1.2   LIMITATIONS OF THE FDTD METHOD

While uncovering a historical perspective of the FDTD method, the previous section presented various strengths of the FDTD method. Of course, the method is not without its weaknesses, and it

Table 1.1: Time-line of significant events in the history of the FDTD method. *Continues.*

Year	Event
1966	Kane Yee's seminal paper [1].
1975	Taflove and Brodwin first to apply Yee's method to EM scattering and biological heating. Correct stability analysis is reported [2, 3].
1977	Holland introduces *THREDE* – first software based on Yee's algorithm for transient EMP simulation [4].
1977	Weiland pioneers the Finite-Integration Technique [16].
1980	Taflove first to apply the acronym FDTD [17].
1981	Holland and Simpson develop a sub cell model to model thin wires [13].
1981	Mur introduces first second-order absorbing boundary for the FDTD method [6].
1981	Gilbert and Holland propose thin-cell sub-cell model [18].
1982-1983	Umashankar and Taflove introduce the TF/SF plane-wave boundary and the near-field to far-field transformation in 2 and 3 dimensions [19, 20].
1983	Holland derives first non orthogonal FDTD (N-FDTD) method [21].
1986	Higdon's Absorbing Boundary Operators [7].
1987	Umashankar, *et al.*, develop sub-cell thin-wire model for FDTD [22].
1988	Sullivan, Gandhi and Taflove – first full human model via the FDTD method [23].
1989	Chu and Chadhuri apply FDTD to optical structures [24].
1990	Sheen, *et al.*, compute the scattering-parameters of printed circuits using the FDTD method [25].
1990	Shankar, *et al.*, and Madsen and Ziolkowski introduce Finite-Volume Time-Domain (FVTD) methods [26, 27].
1990	Sano and Shibata are the first to apply FDTD to opto-electronics [28].
1990-1991	Luebbers, *et al.*, and Joseph, *et al.*, develop stable and accurate FDTD models for frequency-dependent linear dispersive media [29-31].
1991	Maloney, *et al.*, apply FDTD to antenna modeling [32].
1992	Computer Simulation Technology (CST) markets the first FDTD based commercial software - MAFIA (based on the finite-integration technique (FIT) form of FDTD) [33].
1992	Maloney and Smith develop FDTD method to treat frequency dependent surface impedance boundary conditions and thin-material sheets [34, 35].
1992	Goorjian, *et al.*, apply FDTD to solve wave propagation in non-linear media [36].
1992	Jurgens, *et al.*, introduce the Conformal Patch FDTD method [37].
1992	Betz and Mittra introduce an ABC that absorbs evanescent waves [9].
1992-1993	Sui, *et al.*, and Toland, *et al*, first to model non-linear circuit devices within the FDTD method [38, 39].
1993	Schneider and Hudson apply FDTD to anisotropic media [40].
1993	Kunz and Luebbers publish the first text book on FDTD [41].
1994	Luebbers, Langdon and Penney form Remcom, Inc., and market XFDTD [42].
1994	Thomas, *et al.*, couple FDTD model with SPICE for high-speed circuit simulation [43].

Table 1.2: *Continued.* Time-line of significant events in the history of the FDTD method.	
1994	J.-P. Berenger pioneers the Perfectly Matched Layer (PML) absorbing media for simulating unbounded regions with the FDTD method [10].
1994	Chew and Weedon derive a stretched-coordinate form of the PML [44].
1995	FDTD algorithm for massively parallel computers [45].
1995	Madsen introduces the discrete surface integral (DSI) method applying an FDTD-like algorithm to unstructured grids [46].
1995	A. Taflove publishes text book on FDTD (2nd edition in 2000, and 3rd in 2005) [47].
1995-1996	Sacks, Kingsland, Lee, and Lee and Gedney develop an anisotropic media form of the PML [48-50].
1996	Bourgeois and Smith develop a full 3D FDTD model of a ground penetrating radar [51].
1996	Dunn, *et al.*, apply FDTD to study the scattering of light by biological cells.
1997	Dey-Mittra conformal-FDTD method for PEC boundaries [52].
1997	Hagness, *et al,* apply FDTD in the area of photonics and model an optical directional coupler [53, 54].
1998	Roden, *et al.*, develops a stable FDTD analysis of periodic structures with oblique incidence [55].
1998	Hagness, *et al.*, apply FDTD method for microwave imaging for breast cancer detection [56, 57].
2000	Zheng, Chen and Zhang develop the first unconditionally stable 3D alternating direction implicit (ADI) FDTD method [58].
2000	Roden and Gedney develop the Convolutional PML method for the complex-frequency shifted PML parameters [59].
2001	Teixeira, *et al.*, introduce a conformal PML for general coordinate frames [60].
2001	Ziolkowski and Heyman develop FDTD models for negative metamaterials [61].
2002	Chavannes develops stable multi-nested subgrid modeling method [62].
2004	Akyurtlu and Werner develop an FDTD model for bi-anisotropic media [63].
2004	Chang and Taflove combine FDTD with a quantum mechanical atomic model to simulate a four level atomic system [64].
2005	FDTD modeling of photonic nanojets from nanoparticles (Li, *et al*) [65].
2006-2007	Ong, *et al.*, use FDTD to simulate solar cells with nanotube array cathodes [66, 67].
2007	Zhao, *et al.*, use FDTD to model "spatially" dispersive media [68].
2008, 2009	Using the FDTD method, Kong, *et al.*, show that subwavelength pits in a metal substrate can be used for high density optical data storage [69, 70].
2010	Argyropoulos, *et al.*, use FDTD to model the "optical black hole" [71].

is certainly necessary to underline what these are. That is, it is important to understand the limitations of the FDTD method in order to determine when the method is most appropriately applied.

One weakness of the FDTD method is that it requires a full discretization of the electric and magnetic fields throughout the entire volume domain. There are many instances when the FDTD method is forced to model a significant amount of "white-space," i.e., open space with no inhomogeneities. An example could be the electromagnetic scattering of a cluster of perfectly conducting spheres. The white space would be the volume inside each sphere, as well as the region in-between the spheres and the region separating the spheres and the absorbing boundary. The wider the separation between the spheres, the larger the percentage of white space. Another example is a printed circuit in a layered media where the traces being modeled constitute a very small percentage of the volume.

The fact that the FDTD method is fully explicit can also be a weakness. While this can be considered a strength since a linear system of equations is not required to be solved, it can be a weakness if the time-step becomes inordinately small. The time-step is limited by the smallest geometric feature in the model. As a consequence, models with fine, electrically small geometric features can have very small time-steps. This can result in a large number of time-iterations. This poses challenges when applying the FDTD method to large problems with fine geometric features that must be modeled with high fidelity.

The FDTD method provides a broad-band simulation. As a consequence, a broad frequency response can be analyzed via a single simulation. Again, in many circumstances, this can be considered a strength. However, there are instances when only a narrow band response is desired. As a consequence, a frequency domain simulation may be more efficient. Furthermore, when modeling complex materials, if the material constants are only known over a narrow frequency band, this characterization must be approximated over a broadband within the FDTD method. This could be as simple as the loss tangent of a substrate, or the effective material properties of a metamaterial.

The FDTD method can also be challenged when the system under test has a very high Q. Consequently, the time-domain simulation may take a very long time to reach a steady state. Often, this is due to narrow band resonances that slowly decay (if at all). Thus, very long simulation times can result. For some problems, this can be mitigated by using methods such as the Generalized Pencil of Functions [85] to either quantify the resonances, or to extrapolate the signal. In other cases, a frequency-domain simulation about the resonant frequencies may be more expedient.

The FDTD method is also restricted by orthogonal gridding. This can be overcome by the use of subcell modeling techniques as well as sub-gridding and non-uniform grid methods. However, the local fields can still lose accuracy. Also, with methods such as subcell modeling, there is a level of uncertainty as to the exact shape of the local boundary. Thus, it is important to understand the level of accuracy of the near fields that can be achieved by subcell models for the geometry at hand.

Higher-order FDTD methods have not shared the success of higher-order algorithms posed using other methodologies (such as discontinuous Galerkin methods [86, 87]). Most higher-order FDTD methods posed to date are based on an extended stencil – namely, they are based on high-order difference approximations that employ points that span multiple grid cells. This makes it difficult to model simple geometries that involve jump discontinuities in the materials. These methods are also challenged by structures with fine geometric detail.

1.3 ALTERNATE SOLUTION METHODS

It is important to understand that there are a number of other computational methods that have been developed for the simulation of electromagnetic waves. The most popular methods are the method of moments (MoM) [88] and the finite-element method (FEM) [89]. Both methods are typically implemented in the frequency domain. However, time-domain variants are also available. When simulating problems with piece-wise homogeneous materials, the MoM only requires surface discretizations of currents on conducting boundaries, or on boundaries separating different material regions. Therefore, the MoM has a reduced number of unknowns as compared to methods that require a volume discretization. The tradeoff is that the MoM requires the solution of a dense linear system of equations with dimension commensurate to the number of unknowns. The solution of the linear system can be accelerated via the use of fast solution algorithms using either iterative [90] or direct solution methodologies [91]. The method of moments can also be efficiently applied in a layered media. The layered media can be explicitly accounted for via the Green's function [92]. Consequently, only geometries placed within the layered media need be discretized.

The finite-element method (FEM) requires a volume discretization, similar to the FDTD method. However, it is based on unstructured meshing (finite elements) that can accurately resolve fine geometric details. Local basis functions are used to expand the field over each finite element. High-order methods are easily realized. Furthermore, local adaptive refinement of the mesh is commonly employed, realizing controllable accuracy. Similar to the FDTD method, the FEM method also inherently handles inhomogeneous and complex materials. The FEM method requires the solution of a global sparse linear system of equations. Fast solution methods are still not mature for the FEM method. Thus, iterative solutions are commonly used to solve these linear systems of equations for large problems. Such solutions rely on strong preconditioners to be efficient.

Similar to the FDTD method, the MoM and FEM have their strengths and weaknesses. However, it is quite fortunate that in most cases the methods are complementary such that where one method flounders, another excels. For example, the MoM is typically the most efficient method when simulating the electromagnetic interactions involving conducting objects in an unbounded homogeneous medium. The unknowns are restricted to the surface of the conductors. With the use of fast algorithms, an $O(N \log N)$ algorithm can be realized, where N is proportional to the surface area of the conductor, rather than the full volume. The MoM also excels when modeling circuits printed in a layered medium for the same reason. However, if the circuit traces are very densely packed, and/or there are non-planar inhomogeneous regions that occupy a non-negligible fraction of the volume, then the FDTD method can likely be a more efficient method.

The bottom line is that there is not a single computational method that is best suited for all problems. Rather, based on their strengths and weaknesses, each method is better suited for specific classes of problems. Thus, based on the problem being analyzed, the engineer should choose the best method available.

1.4   FDTD SOFTWARE

There are approximately 30 commercial software packages based on the FDTD method that are currently available. The first company to publically market an FDTD software package was Computer Simulation Technology (CST) (founded in 1992), which originally marketed their MAFIA code which was based on the finite-integration technique (FIT) form of the FDTD method. CST's current product is Microwave Studio® [33]. Microwave Studio is geared towards microwaves and RF, signal integrity, and EMC/EMI applications. Remcom software was founded soon thereafter by Ray Luebbers, et al., at Penn State University, in 1994. Remcom currently markets the XFdtd™ software [42], which is also geared towards microwave and RF applications.

Newer competitors in the microwaves and RF market include 2COMU's GEMS™ [93], Speag's SEMCAD X™ [94], Agilent's EMA3D™ [95], Empire's XCcel™ [96], QuickWave™ [97], and Accelerware Corp.'s GPU accelerated software [98]. A number of companies also market FDTD software that is specialized to photonics. This would include Lumericals' FDTD solutions 7.0™ [99], Rsoft's FullWAVE™ [100], Optiwave's OptiFDTD™ [101], NLCSTR's Sim3D_Max™ [102], EM Explorer Studio™ [103], and EM Photonics FastFDTD™ [104].

The number of companies offering FDTD software and FDTD consulting is continuing to grow. This trend will certainly continue as the application areas continue to expand. Of course, there are a plethora of research codes that have also been developed. Some of these have been made public-domain. A partial list can be found by searching for FDTD on Wikipedia.

1.5   OUTLINE TO THE REMAINDER OF THE TEXT

The purpose of this book is to provide a thorough introduction to the Finite-Difference Time-Domain (FDTD) method. The intention is to lay a foundation for the basic FDTD algorithm, including the central difference approximations, explicit time-domain solution methods, boundary conditions, stability, and sources of error, as well as to provide an introduction into more advanced concepts such as absorbing boundary conditions, subcell modeling techniques, and post processing methods. To this end, the book is outlined as follows.

Chapter 2 presents the FDTD solution of the one-dimensional transmission line equations. The intention of this chapter is to provide an introduction of a number of foundational concepts of the FDTD method by applying it to a very simple set of one-dimensional scalar, coupled, first-order differential equations. Within this context, central difference approximations are derived. It is shown that properly staggering the line voltages and currents in space and time will lead to a second-order accurate approximation of the transmission-line equations. The derivation of an explicit recursive update scheme is also provided. The stability of the difference equations is quantified. Also, sources of error in the discrete approximation are unveiled. Finally, simple source and load conditions are derived.

Given the foundation of the one-dimensional problem, the text jumps directly to the three-dimensional Maxwell's equations in Chapter 3. Yee's staggered grid approximation of the fields is presented. It is shown that this discretization directly enables a second-order accurate approximation of Maxwell's coupled curl equations via central differencing. It is also shown that Gauss's laws are inherently satisfied in the discrete space. Similar to Chapter 2, a stability and error analysis of the three-dimensional problem is presented. The treatment of lossy, inhomogeneous, and dispersive materials is also presented in Chapter 3. Finally, more general formulations based on the finite-integration technique and non-uniform gridding are also presented in Chapter 3.

Chapter 4 focuses on transient sources and excitations. A discussion on the properties of the transient signatures and their spectral bandwidths is presented. A number of source types are also presented, including current densities, discrete lumped sources (i.e., Thévenin and Norton equivalent sources), and plane wave excitations.

Chapters 5 and 6 deal with grid truncation techniques for emulating unbounded problem domains. Chapter 5 presents local absorbing boundary conditions (ABC's) based on local differential operators. Chapter 6 presents the perfectly matched layer (PML) absorbing media. Specifically, the Convolutional-PML (CPML) formulation based on a complex frequency shifted (CFS) PML tensor is presented.

Chapter 7 presents a number of subcell modeling techniques. These are more advanced concepts that are important in understanding how modern day commercial FDTD packages simulate more complex structures. Subcell models for thin-wires, narrow slots, and thin-material sheets are presented. Conformal FDTD methods for complex shaped conductor and material boundaries are also introduced.

Chapter 8 presents a number of post-processing methods based on the FDTD simulation. Network parameterizations, including port parameter calculations, port impedances, admittance-parameters, and scattering parameters are detailed. Also, an efficient near-field to far-field transformation is also presented. Antenna gain and scattering cross-section computations are also presented.

Finally, two appendices are provided that focus solely on efficient implementations of the FDTD method using a high-level programming language. Appendix A presents a computer implementation of the one-dimensional FDTD solution of the transmission-line equations. This is aimed at being a entry level programming exercise for the FDTD method. Appendix B deals strictly with the three-dimensional FDTD solution of Maxwell's equations. Appendix B starts with a top-level design of an FDTD code, and then proceeds to detail the various blocks of the design. Significant attention is given to efficient implementations of the FDTD algorithm. Most importantly, a unified algorithmic approach that allows one to model PML, dispersive materials, non-uniform grids, conformal algorithms, etc., within a single code is offered.

Finally, it is hoped that the remainder of this text provides the reader with a sufficient foundational knowledge to be an effective user or developer of the FDTD method. While the material cannot be exhaustive, references are provided throughout that hopefully supplies the reader with sufficient resource to go to levels beyond the content of this text.

REFERENCES

[1] K. S. Yee, "Numerical solution of initial boundary value problems involving Maxwell's equations in isotropic media," *IEEE Transactions on Antennas and Propagation,* vol. AP14, pp. 302–307, 1966. DOI: 10.1109/TAP.1966.1138693 1

[2] A. Taflove and M. E. Brodwin, "Numerical-solution of steady-state electromagnetic scattering problems using time-dependent Maxwell's equations," *IEEE Transactions on Microwave Theory and Techniques,* vol. 23, pp. 623–630, 1975. DOI: 10.1109/TMTT.1975.1128640 1

[3] A. Taflove and M. E. Brodwin, "Computation of electromagnetic-fields and induced temperatures within a model of a microwave-irradiated human eye," *IEEE Transactions on Microwave Theory and Techniques,* vol. 23, pp. 888–896, 1975. DOI: 10.1109/TMTT.1975.1128708 1

[4] R. Holland, "THREDE - Free-field EMP coupling and scattering code," *IEEE Transactions on Nuclear Science,* vol. 24, pp. 2416–2421, 1977. DOI: 10.1109/TNS.1977.4329229 1

[5] C. L. Longmire, "State of the art in IEMP and SGEMP calculations," *IEEE Transactions on Nuclear Science,* vol. NS-22, pp. 2340–2344, 1975. DOI: 10.1109/TNS.1975.4328130 1

[6] G. Mur, "Absorbing boundary-conditions for the finite-difference approximation of the time-domain electromagnetic-field equations," *IEEE Transactions on Electromagnetic Compatibility,* vol. 23, pp. 377–382, 1981. DOI: 10.1109/TEMC.1981.303970 2

[7] R. L. Higdon, "Absorbing boundary-conditions for difference approximations to the multi-dimensional wave-equation " *Mathematics of Computation,* vol. 47, pp. 437–459, Oct 1986. DOI: 10.1090/S0025-5718-1986-0856696-4 2

[8] R. L. Higdon, "Numerical absorbing boundary-conditions for the wave-equation," *Mathematics of Computation,* vol. 49, pp. 65–90, Jul 1987. DOI: 10.1090/S0025-5718-1987-0890254-1 2

[9] V. Betz and R. Mittra, "Comparison and evaluation of boundary conditions for the absorption of guided waves in an FDTD simulation," *IEEE Microwave and Guided Wave Letters,* vol. 2, pp. 499–501, 1992. DOI: 10.1109/75.173408 2

[10] J.-P. Berenger, "A perfectly matched layer for the absorption of electromagnetic waves," *Journal of Computational Physics,* vol. 114, pp. 195–200, 1994. DOI: 10.1006/jcph.1994.1159 2

[11] J. P. Berenger, *Perfect Matched Layer (PML) for Computational Electromagnetics:* Morgan and Claypool Publishers, 2007. DOI: 10.2200/S00030ED1V01Y200605CEM008 2

12 REFERENCES

[12] F. L. Teixeira, "Time-domain finite-difference and finite-element methods for Maxwell equations in complex media," *IEEE Transactions on Antennas and Propagation*, vol. 56, pp. 2150–2166, Aug 2008. DOI: 10.1109/TAP.2008.926767 2

[13] R. Holland and L. Simpson, "Finite-difference analysis of EMP coupling to thin struts and wires," *IEEE Transactions on Electromagnetic Compatibility*, vol. 23, pp. 88–97, 1981. DOI: 10.1109/TEMC.1981.303899 2

[14] S. Benkler, N. Chavannes, and N. Kuster, "A new 3-D conformal PEC FDTD scheme with user-defined geometric precision and derived stability criterion," *IEEE Transactions on Antennas and Propagation*, vol. 54, pp. 1843–1849, Jun 2006. DOI: 10.1109/TAP.2006.875909 3, 4

[15] *SEMCAD X, Reference Manual for the SEMCAD X simulation platform for electromagnetic compatibility, antenna design and dosimetry*, ver. 10.0 ed.: SPEAG - Schmid & Fanner Engineering AG, EIGER, 2005. 3

[16] T. Weiland, "Discretization method for solution of Maxwell's equations for 6-component fields," *AEU-International Journal of Electronics and Communications*, vol. 31, pp. 116–120, 1977.

[17] A. Taflove, "Application of the finite-difference time-domain method to sinusoidal steady-state electromagnetic-penetration problems," *IEEE Transactions on Electromagnetic Compatibility*, vol. 22, pp. 191–202, 1980. DOI: 10.1109/TEMC.1980.303879

[18] J. Gilbert and R. Holland, "Implementation of the thin-slot formalism in the finite difference EMP code THREADII," *IEEE Transactions on Nuclear Science*, vol. 28, pp. 4269–4274, 1981. DOI: 10.1109/TNS.1981.4335711

[19] A. Taflove and K. Umashankar, "Radar cross-section of general 3-dimensional scatterers," *IEEE Transactions on Electromagnetic Compatibility*, vol. 25, pp. 433–440, 1983. DOI: 10.1109/TEMC.1983.304133

[20] K. Umashankar and A. Taflove, "A novel method to analyze electromagnetic scattering of complex objects," *IEEE Transactions on Electromagnetic Compatibility*, vol. 24, pp. 397–405, 1982. DOI: 10.1109/TEMC.1982.304054

[21] R. Holland, "Finite-difference solution of Maxwells equations in generalized nonorthogonal coordinates," *IEEE Transactions on Nuclear Science*, vol. 30, pp. 4589–4591, 1983. DOI: 10.1109/TNS.1983.4333176

[22] K. R. Umashankar, A. Taflove, and B. Beker, "Calculation and experimental validation of induced currents on coupled wires in an arbitrary shaped cavity," *IEEE Transactions on Antennas and Propagation*, vol. 35, pp. 1248–1257, Nov 1987. DOI: 10.1109/TAP.1987.1144000

[23] D. M. Sullivan, O. P. Gandhi, and A. Taflove, "Use of the finite-difference time-domain method for calculating EM absorption in man models," *IEEE Transactions on Biomedical Engineering,* vol. 35, pp. 179–186, 1988. DOI: 10.1109/10.1360

[24] S. T. Chu and S. K. Chaudhuri, "A finite-difference time-domain method for the design and analysis of guided-wave optical structures," *Journal Of Lightwave Technology,* vol. 7, pp. 2033–2038, Dec 1989. DOI: 10.1109/50.41625 4

[25] D. M. Sheen, S. M. Ali, M. D. Abouzahra, and J. A. Kong, "Application of the three-dimensional finite-difference time-domain method to the analysis of planar microstrip circuits," *IEEE Transactions on Microwave Theory and Techniques,* vol. 38, pp. 849–857, July 1990. DOI: 10.1109/22.55775

[26] N. K. Madsen and R. W. Ziolkowski, "A 3-dimensional modified finite volume technique for Maxwell's equations," *Electromagnetics,* vol. 10, pp. 147–161, 1990. DOI: 10.1080/02726349008908233

[27] V. Shankar, A. H. Mohammadian, and W. F. Hall, "A time-domain, finite-volume treatment for the Maxwell equations," *Electromagnetics,* vol. 10, pp. 127–145, 1990. DOI: 10.1080/02726349008908232

[28] E. Sano and T. Shibata, "Fullwave analysis of picosecond photoconductive switches," *IEEE Journal Of Quantum Electronics,* vol. 26, pp. 372–377, Feb 1990. DOI: 10.1109/3.44970

[29] R. Luebbers, F. P. Hunsberger, K. S. Kunz, R. B. Standler, and M. Schneider, "A Frequency-Dependent Finite-Difference Time-Domain Formulation for Dispersive Materials," *IEEE Transactions on Electromagnetic Compatibility,* vol. 32, pp. 222–227, 1990. DOI: 10.1109/15.57116

[30] R. J. Luebbers and F. Hunsberger, "FDTD for Nth-Order Dispersive Media," *IEEE Transactions on Antennas and Propagation,* vol. 40, pp. 1297–1301, 1992. DOI: 10.1109/8.202707

[31] R. M. Joseph, S. C. Hagness, and A. Taflove, "Direct time integration of Maxwell equations in linear dispersive media with absorption for scattering and propagation of femtosecond electromagnetic pulses," *Optics Letters,* vol. 16, pp. 1412–1414, Sep 1991. DOI: 10.1364/OL.16.001412

[32] J. G. Maloney, G. S. Smith, and W. R. Scott, "Accurate computation of the radiation from simple antennas using the finite-difference time-domain method," *IEEE Transactions on Antennas and Propagation,* vol. 38, pp. 1059–1068, Jul 1990. DOI: 10.1109/8.55618

[33] http://www.cst.com/. 9

14 REFERENCES

[34] J. G. Maloney and G. S. Smith, "The Use of Surface Impedance Concepts in the Finite-Difference Time-Domain Method," *IEEE Transactions on Antennas and Propagation,* vol. 40, pp. 38–48, 1992. DOI: 10.1109/8.123351

[35] J. G. Maloney and G. S. Smith, "The efficient modeling of thin material sheets in the finite-difference time-domain (FDTD) method," *IEEE Transactions on Antennas and Propagation,* vol. 40, pp. 323–330, Mar 1992. DOI: 10.1109/8.135475

[36] P. M. Goorjian and A. Taflove, "Direct time integration of Maxwell's equations in nonlinear dispersive media for propagation and scattering of femtosecond electromagnetic solitons," *Optics Letters,* vol. 17, pp. 180–182, 1992. DOI: 10.1364/OL.17.000180

[37] T. G. Jurgens, A. Taflove, K. R. Umashankar, and T. G. Moore, "Finite-difference time-domain modeling of curved surfaces," *IEEE Transactions on Antennas and Propagation,* vol. 40, pp. 357–366, 1992. DOI: 10.1109/8.138836

[38] W. Q. Sui, D. A. Christensen, and C. H. Durney, "Extending the 2-dimensional FDTD method to hybrid electromagnetic systems with active and passive lumped elements," *IEEE Transactions on Microwave Theory and Techniques,* vol. 40, pp. 724–730, Apr 1992. DOI: 10.1109/22.127522

[39] B. Toland, B. Houshmand, and T. Itoh, "Modeling of nonlinear active regions with the FDTD method," *IEEE Microwave and Guided Wave Letters,* vol. 3, pp. 333–335, 1993. DOI: 10.1109/75.244870

[40] J. B. Schneider and S. Hudson, "The finite-difference time-domain method applied to anisotropic material," *IEEE Transactions on Antennas and Propagation,* vol. 41, pp. 994–999, 1993. DOI: 10.1109/8.237636

[41] K. S. Kunz and R. Luebbers, *Finite Difference Time Domain Method for Electromagnetics,* Boca Raton, FL: CRC Press, LLC, 1993.

[42] http://www.remcom.com/. 9

[43] V. A. Thomas, M. E. Jones, M. J. Piket-May, A. Taflove, and E. Harrigan, "The use of SPICE lumped circuits as sub-grid models for FDTD high-speed electronic circuit design," *IEEE Microwave and Guided Wave Letters,* vol. 4, pp. 141–143, 1994. DOI: 10.1109/75.289516

[44] W. C. Chew and W. H. Weedon, "A 3D Perfectly Matched Medium from Modified Maxwells Equations with Stretched Coordinates," *Microwave and Optical Technology Letters,* vol. 7, pp. 599–604, 1994. DOI: 10.1002/mop.4650071304

[45] S. D. Gedney, "Finite-Difference Time-Domain Analysis of Microwave Circuit Devices on High-Performance Vector/Parallel Computers," *IEEE Transactions on Microwave Theory and Techniques,* vol. 43, pp. 2510–2514, 1995. DOI: 10.1109/22.466191

[46] N. K. Madsen, "Divergence preserving discrete surface integral methods for Maxwell's equations using nonorthogonal unstructured grids," *Journal of Computational Physics*, vol. 119, pp. 34–45, 1995. DOI: 10.1006/jcph.1995.1114

[47] A. Taflove, "Computational Electrodynamics: The Finite Difference Time Domain," Boston: Artech House, 1995.

[48] S. D. Gedney, "An anisotropic perfectly matched layer-absorbing medium for the truncation of FDTD lattices," *IEEE Transactions on Antennas and Propagation*, vol. 44, pp. 1630–1639, December 1996. DOI: 10.1109/8.546249

[49] S. D. Gedney, "An anisotropic PML absorbing media for the FDTD simulation of fields in lossy and dispersive media," *Electromagnetics*, vol. 16, pp. 399–415, 1996. DOI: 10.1080/02726349608908487

[50] Z. S. Sacks, D. M. Kingsland, R. Lee, and J. F. Lee, "A perfectly matched anisotropic absorber for use as an absorbing boundary condition," *IEEE Transactions on Antennas and Propagation*, vol. 43, pp. 1460–1463, 1995. DOI: 10.1109/8.477075

[51] J. M. Bourgeois and G. S. Smith, "A fully three-dimensional simulation of a ground-penetrating radar: FDTD theory compared with experiment," *IEEE Transactions on Geoscience and Remote Sensing*, vol. 34, pp. 36–44, 1996. DOI: 10.1109/36.481890 4

[52] S. Dey and R. Mittra, "A locally conformal finite-difference time-domain (FDTD) algorithm for modeling three-dimensional perfectly conducting objects," *IEEE Microwave And Guided Wave Letters*, vol. 7, pp. 273–275, Sep 1997. DOI: 10.1109/75.622536

[53] S. C. Hagness, D. Rafizadeh, S. T. Ho, and A. Taflove, "FDTD microcavity simulations: Design and experimental realization of waveguide-coupled single-mode ring and whispering-gallery-mode disk resonators," *Journal Of Lightwave Technology*, vol. 15, pp. 2154–2165, Nov 1997. DOI: 10.1109/50.641537

[54] D. Rafizadeh, J. P. Zhang, S. C. Hagness, A. Taflove, K. A. Stair, S. T. Ho, and R. C. Tiberio, "Waveguide-coupled AlGaAs/GaAs microcavity ring and disk resonators with high finesse and Z1.6-nm free spectral range," *Optics Letters*, vol. 22, pp. 1244–1246, Aug 1997. DOI: 10.1364/OL.22.001244

[55] J. A. Roden, S. D. Gedney, M. P. Kesler, J. G. Maloney, and P. H. Harms, "Time-domain analysis of periodic structures at oblique incidence: Orthogonal and nonorthogonal FDTD implementations," *IEEE Transactions on Microwave Theory and Techniques*, vol. 46, pp. 420–427, 1998. DOI: 10.1109/22.664143

[56] S. C. Hagness, A. Taflove, and J. E. Bridges, "Two-dimensional FDTD analysis of a pulsed microwave confocal system for breast cancer detection: Fixed-focus and antenna-array sensors," *IEEE Transactions on Biomedical Engineering,* vol. 45, pp. 1470–1479, Dec 1998. DOI: 10.1109/10.730440

[57] S. C. Hagness, A. Taflove, and J. E. Bridges, "Three-dimensional FDTD analysis of a pulsed microwave confocal system for breast cancer detection: Design of an antenna-array element," *IEEE Transactions on Antennas and Propagation,* vol. 47, pp. 783–791, May 1999. DOI: 10.1109/8.774131

[58] F. H. Zheng, Z. Z. Chen, and J. Z. Zhang, "Toward the development of a three-dimensional unconditionally stable finite-difference time-domain method," *IEEE Transactions on Microwave Theory and Techniques,* vol. 48, pp. 1550–1558, 2000. DOI: 10.1109/22.869007

[59] J. A. Roden and S. D. Gedney, "Convolutional PML (CPML): An Efficient FDTD Implementation of the CFS-PML for Arbitrary Media," *Microwave and Optical Technology Letters,* vol. 27, pp. 334–339, December 5 2000. DOI: 10.1002/1098-2760(20001205)27:5%3C334::AID-MOP14%3E3.3.CO;2-1

[60] F. L. Teixeira, K. P. Hwang, W. C. Chew, and J. M. Jin, "Conformal PML-FDTD schemes for electromagnetic field simulations: A dynamic stability study," *IEEE Transactions On Antennas And Propagation,* vol. 49, pp. 902–907, Jun 2001. DOI: 10.1109/8.931147

[61] R. W. Ziolkowski and E. Heyman, "Wave propagation in media having negative permittivity and permeability," *Physical Review E,* vol. 64, 2001. DOI: 10.1103/PhysRevE.64.056625

[62] N. Chavannes, "Local mesh refinement algorithms for enhanced modeling capabilities in the FDTD method," Zurich, Switzerland: Swiss Federal Institute of Technology (ETH), 2002.

[63] A. Akyurtlu and D. H. Werner, "Modeling of transverse propagation through a uniaxial bianisotropic medium using the finite-difference time-domain technique," *IEEE Transactions on Antennas and Propagation,* vol. 52, pp. 3273–3279, Dec 2004. DOI: 10.1109/TAP.2004.836442

[64] S. H. Chang and A. Taflove, "Finite-difference time-domain model of lasing action in a four-level two-electron atomic system," *Optics Express,* vol. 12, pp. 3827–3833, 2004. DOI: 10.1364/OPEX.12.003827

[65] X. Li, Z. G. Chen, A. Taflove, and V. Backman, "Optical analysis of nanoparticles via enhanced backscattering facilitated by 3-D photonic nanojets," *Optics Express,* vol. 13, pp. 526–533, Jan 2005. DOI: 10.1364/OPEX.13.000526

[66] G. K. Mor, O. K. Varghese, M. Paulose, K. Shankar, and C. A. Grimes, "A review on highly ordered, vertically oriented TiO2 nanotube arrays: Fabrication, material properties, and solar

energy applications," *Solar Energy Materials and Solar Cells*, vol. 90, pp. 2011–2075, Sep 2006. DOI: 10.1016/j.solmat.2006.04.007 4

[67] K. G. Ong, O. K. Varghese, G. K. Mor, K. Shankar, and C. A. Grimes, "Application of finite-difference time domain to dye-sensitized solar cells: The effect of nanotube-array negative electrode dimensions on light absorption," *Solar Energy Materials and Solar Cells*, vol. 91, pp. 250–257, Feb 2007. DOI: 10.1016/j.solmat.2006.09.002 4

[68] Y. Zhao, P. A. Belov, and Y. Hao, "Modelling of wave propagation in wire media using spatially dispersive finite-difference time-domain method: Numerical aspects," *IEEE Transactions on Antennas and Propagation*, vol. 55, pp. 1506–1513, 2007. DOI: 10.1109/TAP.2007.897320

[69] S. C. Kong, A. Sahakian, A. Taflove, and V. Backman, "Photonic nanojet-enabled optical data storage," *Optics Express*, vol. 16, pp. 13713–13719, Sep 2008. DOI: 10.1364/OE.16.013713 4

[70] S. C. Kong, A. V. Sahakian, A. Taflove, and V. Backman, "High-Density Optical Data Storage Enabled by the Photonic Nanojet from a Dielectric Microsphere," *Japanese Journal of Applied Physics*, vol. 48, p. 3, Mar 2009. DOI: 10.1143/JJAP.48.03A008

[71] C. Argyropoulos, E. Kallos, and Y. Hao, "FDTD analysis of the optical black hole," *Journal of the Optical Society of America B-Optical Physics*, vol. 27, pp. 2020–2025, Oct. 2010. DOI: 10.1364/JOSAB.27.002020 4

[72] C. Furse, "A survey of phased arrays for medical applications," *Applied Computational Electromagnetics Society Journal*, vol. 21, pp. 365–379, Nov 2006. 4

[73] A. Dunn and R. Richards-Kortum, "Three-dimensional computation of light scattering from cells," *IEEE Journal of Selected Topics in Quantum Electronics*, vol. 2, pp. 898–905, Dec 1996. DOI: 10.1109/2944.577313

[74] A. Wax and V. Backman, *Biomedical Applications of Light Scattering*: McGraw Hill, 2010.

[75] J. D. Shea, P. Kosmas, S. C. Hagness, and B. D. Van Veen, "Three-dimensional microwave imaging of realistic numerical breast phantoms via a multiple-frequency inverse scattering technique," *Medical Physics*, vol. 37, pp. 4210–4226, August 2010. DOI: 10.1118/1.3443569 4

[76] O. P. Gandhi, "Electromagnetic fields: Human safety issues," *Annual Review of Biomedical Engineering*, vol. 4, pp. 211–234, 2002. DOI: 10.1146/annurev.bioeng.4.020702.153447 4

[77] L. Gurel and U. Oguz, "Three-dimensional FDTD modeling of a ground-penetrating radar," *IEEE Transactions on Geoscience and Remote Sensing*, vol. 38, pp. 1513–1521, Jul 2000. DOI: 10.1109/36.851951 4

[78] S. C. Kong and Y. W. Choi, "Finite-Difference Time-Domain (FDTD) Model for Traveling-Wave Photodetectors," *Journal of Computational and Theoretical Nanoscience*, vol. 6, pp. 2380–2387, Nov 2009. DOI: 10.1166/jctn.2009.1292 4

[79] H. G. Park, S. H. Kim, S. H. Kwon, Y. G. Ju, J. K. Yang, J. H. Baek, S. B. Kim, and Y. H. Lee, "Electrically driven single-cell photonic crystal laser," *Science*, vol. 305, pp. 1444–1447, Sep 2004. DOI: 10.1126/science.1100968

[80] R. W. Ziolkowski and J. B. Judkins, "Applications of the nonlinear finite-difference time-domain (NL-FDTD) method to pulse-propagation in nonlinear media - self-focusing and linear-nonlinear interfaces," *Radio Science*, vol. 28, pp. 901–911, Sep-Oct 1993. DOI: 10.1029/93RS01100 4

[81] X. Li, A. Taflove, and V. Backman, "Recent progress in exact and reduced-order modeling of light-scattering properties of complex structures," *IEEE Journal of Selected Topics in Quantum Electronics*, vol. 11, pp. 759–765, 2005. DOI: 10.1109/JSTQE.2005.857691 4

[82] S. Tanev, W. B. Sun, J. Pond, V. V. Tuchin, and V. P. Zharov, "Flow cytometry with gold nanoparticles and their clusters as scattering contrast agents: FDTD simulation of light-cell interaction," *Journal of Biophotonics*, vol. 2, pp. 505–520, Sep 2009. DOI: 10.1002/jbio.200910039 4

[83] E. Ozbay, "Plasmonics: Merging photonics and electronics at nanoscale dimensions," *Science*, vol. 311, pp. 189–193, Jan 2006. DOI: 10.1126/science.1114849 4

[84] J. J. Simpson and A. Taflove, "Electrokinetic effect of the Loma Prieta earthquake calculated by an entire-Earth FDTD solution of Maxwell's equations," *Geophysical Research Letters*, vol. 32, 2005. DOI: 10.1029/2005GL022601 4

[85] T. K. Sarkar and O. Pereira, "Using the matrix pencil method to estimate the parameters of a sum of complex exponentials," *IEEE Antennas and Propagation Magazine*, vol. 37, pp. 48–55, Feb 1995. DOI: 10.1109/74.370583 7

[86] B. Cockburn, G. E. Karniadakis, and C.-W. Shu, *Discontinuous Galerkin Methods: Theory, Computation and Applications* vol. 11. Berlin: Springer-Verlag Telos, 2000. 7

[87] J. S. Hesthaven and T. Warburton, "High-order nodal discontinuous Galerkin methods for the Maxwell eigenvalue problem," *Philosophical Transactions Of The Royal Society Of London Series A-Mathematical Physical And Engineering Sciences*, vol. 362, pp. 493–524, Mar 15 2004. DOI: 10.1098/rsta.2003.1332 7

[88] A. F. Peterson, S. L. Ray, and R. Mittra, *Computational Methods for Electromagnetics*. New York: IEEE Press, 1998. 8

[89] J. M. Jin, *The Finite Element Method in Electromagnetics*, 2nd ed. New York: John Wiley & Sons, Inc., 2002. 8

[90] W. C. Chew, J. M. Jin, E. Michielssen, and J. M. Song, *Fast and Efficient Algorithms in Computational Electromagnetics*: Artech House, Boston, MA, 2001. 8

[91] R. J. Adams, Y. Xu, X. Xu, J. S. Choi, S. D. Gedney, and F. X. Canning, "Modular fast direct electromagnetic analysis using local-global solution modes," *IEEE Transactions on Antennas and Propagation*, vol. 56, pp. 2427–2441, Aug 2008. DOI: 10.1109/TAP.2008.926769 8

[92] K. A. Michalski and J. R. Mosig, "Multilayered media Green's functions in integral equation formulations," *IEEE Transactions on Antennas and Propagation*, vol. 45, pp. 508–519, Mar 1997. DOI: 10.1109/8.558666 8

[93] http://www.2comu.com/. 9

[94] http://www.speag.com/. 9

[95] http://electromagneticapplications.com/ema3d_main.html. 9

[96] http://www.empire.de/. 9

[97] http://www.qwed.com.pl/. 9

[98] http://www.acceleware.com/fdtd-solvers. 9

[99] http://www.lumerical.com/fdtd.php. 9

[100] http://rsoftdesign.com. 9

[101] http://www.optiwave.com/. 9

[102] http://www.nlcstr.com/sim3d.htm. 9

[103] http://www.emexplorer.net/. 9

[104] http://www.emphotonics.com/. 9

CHAPTER 2

1D FDTD Modeling of the Transmission Line Equations

2.1 THE TRANSMISSION LINE EQUATIONS

Transmission lines serve to guide the propagation of electromagnetic energy from a source terminal to a load terminal. Transmission lines can take on many physical forms, including a twisted pair line used for telephony or internet connections, coaxial cables, or any of a number of multi-conductor wave guiding systems [1]. Transmission line modeling is typically a one-dimensional approximation of the physical model, which represents line voltages and currents as a function of the transmission line axis. The time-dependent transmission line equations[1] governing the line voltages and currents are expressed as [1]:

$$\frac{\partial I(x,t)}{\partial x} = -C\frac{\partial V(x,t)}{\partial t}, \quad \frac{\partial V(x,t)}{\partial x} = -L\frac{\partial I(x,t)}{\partial t}, \tag{2.1}$$

where $V(x,t)$ is the line voltage at position x along the transmission line axis at time t, $I(x,t)$ is the line current, and L and C are the per unit length inductance (H/m) and capacitance (F/m), respectively. Note that a lossless transmission line has been assumed for simplicity. The two equations can be combined into the familiar wave equation governing the line voltage:

$$\frac{\partial^2 V(x,t)}{\partial x^2} = LC\frac{\partial^2 V(x,t)}{\partial t^2}. \tag{2.2}$$

A dual wave equation is derived for the line current. This equation has the well-known solution expressed as a superposition of forward and backward traveling waves

$$V(x,t) = A^+ f\left(x - t/\sqrt{LC}\right) + A^- f\left(x + t/\sqrt{LC}\right) \tag{2.3}$$

where $f(*)$ is a general function representing the wave, and $A^{+/-}$ is the amplitude of the forward and the backward traveling waves, respectively. The actual values of f and $A^{+/-}$ are determined by the appropriate initial conditions and boundary conditions at the terminating ends of the transmission line.

---

[1]The transmission line equations are also referred to as the "telegrapher's equations."

2.2   FINITE DIFFERENCE APPROXIMATIONS

The objective of this chapter is to study the numerical solution to the coupled first-order transmission line equations expressed in (2.1). To this end, the first-order space and time derivatives will be approximated via "finite-difference" approximations. If the function $f(x)$ is C^{∞}-continuous, a Taylor series expansion of f about the points $x \pm \Delta x/2$ can be performed [2]:

$$f\left(x \pm \frac{\Delta x}{2}\right) = f(x) \pm \frac{\partial f(x)}{\partial x} \frac{\Delta x}{2} \frac{1}{1!} + \frac{\partial^2 f(x)}{\partial x^2} \left(\frac{\Delta x}{2}\right)^2 \frac{1}{2!} \pm \frac{\partial^3 f(x)}{\partial x^3} \left(\frac{\Delta x}{2}\right)^3 \frac{1}{3!} + \ldots \quad (2.4)$$

where Δx is assumed to be small. We then subtract the two expansions, and normalize the result by Δx, leading to:

$$\frac{f(x + \frac{\Delta x}{2}) - f\left(x - \frac{\Delta x}{2}\right)}{\Delta x} = \frac{\partial f(x)}{\partial x} + \Delta x^2 \frac{\partial^3 f(x)}{\partial x^3} \frac{1}{24} + \ldots \quad (2.5)$$

Note that the terms with even-order derivatives cancel. Rearranging terms, the first-order derivative with respect to x can be expressed as:

$$\frac{\partial f(x)}{\partial x} \approx \frac{f\left(x + \frac{\Delta x}{2}\right) - f\left(x - \frac{\Delta x}{2}\right)}{\Delta x} + O\left(\Delta x^2\right). \quad (2.6)$$

This is known as the "central difference" approximation of the first-order derivative [3]. The trailing term on the right-hand-side represents the leading order error in the approximation. That is, the error will decay as Δx^2 with decreasing Δx. The approximation is thus said to be "second-order accurate."

From (2.5), it is observed that the *amplitude* of the error is proportional to the third-order derivative of f. Thus, the error is impacted by the relative smoothness of f. Consequently, the more rapidly varying f is, the larger the error is, and, consequently, the smaller Δx must be to reduce the error to desired precision. When modeling wave propagation problems, the rate of undulation of a wave is dependent upon the wavelength of the propagating signal. Consequently, a central-difference approximation of the wave will require Δx to be small relative to the smallest wavelength.

The central difference approximation requires discrete samples of f at two points that are centered about x. As seen in the derivation of (2.6) from (2.4), this leads to the cancellation of the terms with even-order derivatives. If the two discrete samples are not centered, the even-order derivative terms will not cancel. Consequently, the approximation will only be first-order accurate.

2.3   EXPLICIT TIME UPDATE SOLUTION

It would appear that the transmission line equations can be approximated to second-order accuracy using central differences. To this end, the transmission line equations of (2.1) are expressed as:

$$\frac{I(x + \Delta x/2, t) - I(x - \Delta x/2, t)}{\Delta x} \approx -C \frac{V(x, t + \Delta t/2) - V(x, t - \Delta t/2)}{\Delta t}, \tag{2.7}$$

$$\frac{V(x + \Delta x/2, t) - V(x - \Delta x/2, t)}{\Delta x} \approx -L \frac{I(x, t + \Delta t/2) - I(x, t - \Delta t/2)}{\Delta t}, \tag{2.8}$$

where Δx is the spacing between discrete spatial samples, and Δt is the spacing between discrete time samples. Both of these equations are second-order accurate at (x, t) in space and time. These approximate equations can be mapped to a discrete space by sampling V and I at discrete locations in space and time at uniformly spaced intervals of Δx and Δt, respectively. However, it is observed from (2.7) and (2.8) that the discrete samples of V and I cannot be co-located in space or time if these two equations are to be consistent. What is observed is that the two equations can be rendered consistent if the discrete samples of V and I are separated in both space and time by $\Delta x/2$ and $\Delta t/2$, respectively. This is illustrated in Fig. 2.1. By staggering the line voltage and current in this manner, second-order accuracy is maintained. Following this approach, the discrete representation for a uniform, staggered time- and space-sampling of V and I is defined as:

$$V_i^{n+\frac{1}{2}} = V\left(i\Delta x, (n + \tfrac{1}{2})\Delta t\right) \tag{2.9}$$

and

$$I_{i+\frac{1}{2}}^{n} = I\left((i + \tfrac{1}{2})\Delta x, n\Delta t\right) \tag{2.10}$$

where i and n are integers. With this sampling, the discrete transmission line equations (2.7) and (2.8) are expressed as:

$$\frac{I_{i+\frac{1}{2}}^{n} - I_{i-\frac{1}{2}}^{n}}{\Delta x} \approx -C \frac{V_i^{n+\frac{1}{2}} - V_i^{n-\frac{1}{2}}}{\Delta t}, \tag{2.11}$$

$$\frac{V_{i+1}^{n+\frac{1}{2}} - V_i^{n+\frac{1}{2}}}{\Delta x} \approx -L \frac{I_{i+\frac{1}{2}}^{n+1} - I_{i+\frac{1}{2}}^{n}}{\Delta t}. \tag{2.12}$$

The stencil of each of these two operators is illustrated in Fig. 2.1.

Now, consider the initial value problem, where $V_i^{n-\frac{1}{2}}$ and $I_{i+\frac{1}{2}}^{n}$ are known at time n for all i.[2] Then from (2.11) and (2.12), a simple recursive relationship can be derived to predict the values of

---

[2]The initial values for V and I at all discrete spatial points must satisfy the transmission line equations. Typically, they are both initialized to 0 for $n = 0$.

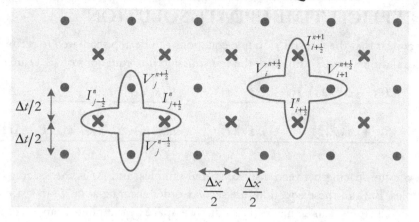

Figure 2.1: Staggered grid sampling of V (circles) and I (x's) in space and time. Also illustrated is the stencil of points use for the central difference approximation of the two transmission-line equations.

$V_i^{n+\frac{1}{2}}$ and $I_{i+\frac{1}{2}}^{n+1}$:

$$V_i^{n+\frac{1}{2}} = V_i^{n-\frac{1}{2}} - \frac{\Delta t}{C \Delta x} \left(I_{i+\frac{1}{2}}^n - I_{i-\frac{1}{2}}^n \right),$$ (2.13)

$$I_{i+\frac{1}{2}}^{n+1} = I_{i+\frac{1}{2}}^n - \frac{\Delta t}{L \Delta x} \left(V_{i+1}^{n+\frac{1}{2}} - V_i^{n+\frac{1}{2}} \right).$$ (2.14)

This recursive formulation can be used to "advance" the line voltages and currents on the transmission-line in time. This recursive scheme is known as an "explicit" difference operator. It is explicit because it is based purely on previous time-values of the line voltage and current, and it only involves local spatial samples. Referring again to Fig. 2.1, it is seen that $I_{i+\frac{1}{2}}^{n+1}$ relies only on $V_{i+1}^{n+\frac{1}{2}}$ and $V_i^{n+\frac{1}{2}}$, which are spatially centered about $I_{i+\frac{1}{2}}^{n+1}$, and are at the previous half time-step, and $I_{i+\frac{1}{2}}^n$ at the previous time-step. The dual holds true for $V_i^{n+\frac{1}{2}}$.

The recursive update equations (2.13) and (2.14) are also referred to as a "leap-frog" time-udpate strategy since the voltages and currents are computed in an alternating manner. Namely, the voltages are updated at time-step $n + \frac{1}{2}$ for all space samples. The currents are then updated at time-step $n+1$ for all space samples, and so on.

The accuracy of this time-update procedure based on the space sampling of the line currents and voltages, the bandwidth of the pulse being propagated along the line and the discrete time sampling will be explored in Section 2.4. Another important issue is that the time-sampling cannot be arbitrarily chosen. Rather, it must satisfy a specific stability criterion. This is the topic of Section 2.5.

2.4   NUMERICAL DISPERSION

The central difference approximations of the differential operators introduces inherent numerical error. It is important to characterize the error of the discrete solution in order to predict with confidence time and space discretizations that will keep the error below a desired minimum level. One way this can be done is to propagate a known signal along the transmission line and predict the error as a function of the space and time sampling versus the pulse bandwidth. Perhaps the simplest known signal to study is a monochromatic plane wave that is propagating on a matched transmission line. This is what is studied in this section.

To study the wave propagation along the discretized transmission line, (2.13) and (2.14) are first combined into a discrete-form of the wave equation. For example, $V_{i+1}^{n+\frac{1}{2}}$ and $V_i^{n+\frac{1}{2}}$ in (2.14) are substituted with the expressions in (2.13), leading to

$$I_{i+\frac{1}{2}}^{n+1} = I_{i+\frac{1}{2}}^{n} - \frac{\Delta t}{L\Delta x}\left[\left(V_{i+1}^{n-\frac{1}{2}} - \frac{\Delta t}{C\Delta x}\left(I_{i+\frac{3}{2}}^{n} - I_{i+\frac{1}{2}}^{n}\right)\right) - \left(V_i^{n-\frac{1}{2}} - \frac{\Delta t}{C\Delta x}\left(I_{i+\frac{1}{2}}^{n} - I_{i-\frac{1}{2}}^{n}\right)\right)\right]. \quad (2.15)$$

Noting that

$$V_{i+1}^{n-\frac{1}{2}} - V_i^{n-\frac{1}{2}} = -\frac{L\Delta x}{\Delta t}\left[I_{i+\frac{1}{2}}^{n} - I_{i+\frac{1}{2}}^{n-1}\right], \quad (2.16)$$

(2.15) can be rewritten as:

$$I_{i+\frac{1}{2}}^{n+1} = 2I_{i+\frac{1}{2}}^{n} - I_{i+\frac{1}{2}}^{n-1} + \frac{(c\Delta t)^2}{\Delta x^2}\left[I_{i+\frac{3}{2}}^{n} - 2I_{i+\frac{1}{2}}^{n} + I_{i-\frac{1}{2}}^{n}\right], \quad (2.17)$$

where

$$c = \frac{1}{\sqrt{LC}} \quad (2.18)$$

is the wave-propagation speed on the transmission line. This represents a discrete form of the wave equation. A dual equation can be derived for $V_i^{n+\frac{1}{2}}$. Either equation can be expressed as a general second-order difference operator:

$$f_i^{n+1} = 2f_i^{n} - f_i^{n-1} + \frac{(c\Delta t)^2}{\Delta x^2}\left[f_{i+1}^{n} - 2f_i^{n} + f_{i-1}^{n}\right] \quad (2.19)$$

where f represents either the line voltage or current.

Now, let f be expressed as a monochromatic wave, propagating along the transmission line. In the continuous space

$$f(x, t) = e^{j(\omega t - kx)} \quad (2.20)$$

where ω is the radial frequency ($\omega = 2\pi f$), f is the frequency in Hz, and k is the wave-number, defined as:

$$k = \frac{\omega}{c} = \frac{2\pi f}{c} = \frac{2\pi}{\lambda}, \tag{2.21}$$

and λ is the wavelength [4]. In the discrete space,

$$f(i\Delta x, n\Delta t) = f_i^n = e^{j(\omega n \Delta t - \tilde{k} i \Delta x)} \tag{2.22}$$

where $\tilde{k}$ is the numerical wave-number of the discrete space. The discrete plane wave is then inserted into (2.19), leading to:

$$e^{j(\omega(n+1)\Delta t - \tilde{k} i \Delta x)} = (c\Delta t)^2 \left[\frac{e^{j(\omega n \Delta t - \tilde{k}(i+1)\Delta x)} - 2e^{j(\omega n \Delta t - \tilde{k} i \Delta x)} + e^{j(\omega n \Delta t - \tilde{k}(i-1)\Delta x)}}{\Delta x^2} \right].$$
$$+ 2e^{j(\omega n \Delta t - \tilde{k} i \Delta x)} - e^{j(\omega(n-1)\Delta t - \tilde{k} i \Delta x)} \tag{2.23}$$

Factoring out $e^{j(\omega n \Delta t - \tilde{k} i \Delta x)}$, grouping terms, and applying Euler's law [2], this reduces to:

$$\cos(\omega \Delta t) = \frac{(c\Delta t)^2}{\Delta x^2} \left[\cos(\tilde{k} \Delta x) - 1 \right] + 1. \tag{2.24}$$

This equation represents the "dispersion relation" of the discrete wave equation.

To study the behavior of this relationship, initially assume space and time sampling in the limit that $\Delta x \to 0$ and $\Delta t \to 0$. Assuming a Taylor series expansions for the cosine functions [2], (2.24) can be expressed as:

$$1 - \frac{(\omega \Delta t)^2}{2} \cong \frac{(c\Delta t)^2}{\Delta x^2} \left[1 - \frac{(\tilde{k} \Delta x)^2}{2} - 1 \right] + 1. \tag{2.25}$$

Cancelling terms, this reduces to:

$$\omega^2 = c^2 \tilde{k}^2 \Rightarrow \quad \tilde{k} = \pm \frac{\omega}{c} \tag{2.26}$$

which is the exact dispersion relation of (2.21). Consequently, the finite-difference time-domain approximation in (2.13) and (2.14) converge to the exact solution in the limit of diminishing space and time sampling.

In general, the numerical wave number is derived from (2.24) as:

$$\tilde{k} = \frac{1}{\Delta x} \cos^{-1} \left[\frac{\Delta x^2}{(c\Delta t)^2} (\cos(\omega \Delta t) - 1) + 1 \right]. \tag{2.27}$$

If one chooses $c\Delta t = \Delta x$, then (2.27) reduces to

$$\tilde{k} = \frac{1}{\Delta x} \cos^{-1} \left[\left(\cos(\frac{\omega \Delta x}{c}) - 1 \right) + 1 \right] = \frac{\omega}{c}, \tag{2.28}$$

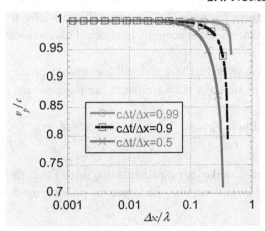

Figure 2.2: Numerical phase velocity normalized by the exact wave speed versus $\Delta x/\lambda$.

which is the exact solution! With this choice, the errors due to the central difference approximations in time and space exactly cancel. Note that this is independent of the step size and the frequency. For this reason, this time-step has been referred to as the "magic time-step" [5]. This may seem like a phenomenal discovery. However, this magic time-step is limited to the scope of a single one-dimensional lossless transmission line with constant line parameters. Nevertheless, it will provide some useful insights.

A more general choice would be to let $c\Delta t < \Delta x$, which is within the stable region (c.f., Section 2.5). In this situation, the numerical wave number $\tilde{k} \neq k$, and as a consequence, there will be error in the solution. The error can also be measured by the effective numerical phase velocity of the discrete wave, defined as [4]:

$$\tilde{v}_p = \frac{\omega}{\tilde{k}} = \frac{2\pi c}{\cos^{-1}\left[\frac{\Delta x^2}{(c\Delta t)^2}\left(\cos(\omega\Delta t) - 1\right) + 1\right]} \frac{\Delta x}{\lambda} \tag{2.29}$$

where the relationship $\omega = 2\pi c/\lambda$ has been used. It is observed that the numerical phase velocity is frequency dependent, and, consequently, the wave is said to be dispersive. It is also observed that the error is strongly related to the number of space samples per wavelength. A study of the numerical phase velocity $\tilde{v}_p$ normalized by the exact wave speed c is presented by Fig. 2.2. The numerical phase velocity is computed via (2.29) as a function of the spatial sampling (Δx) per wavelength for different time-steps ranging from $c\Delta t = 0.99\Delta x$ to $c\Delta t = 0.5\Delta x$. What is observed is that the error is reduced as the time-step is increased. This is typical, even in higher-dimensions. What is also observed is that the error also increases with increasing $\Delta x/\lambda$. This dictates that the higher the frequency spectrum of the signal being propagated, the smaller the space step size must be. At high frequencies, when the argument of the inverse cosine function exceeds 1.0, the phase velocity becomes complex. At these frequencies, the wave will artificially attenuate. The curves in Fig. 2.2 stop just prior to the phase velocity becoming complex. This cross-over point occurs near, but below, the Nyquist sampling rate.

Due to the dispersive error, the signal will accumulate phase error as it propagates along the transmission line. The phase error per wavelength can be measured as:

$$\frac{\text{Phase Error}}{\lambda} = \frac{(\tilde{k} - k)}{k} \cdot \frac{360°}{\lambda} = \left(\frac{c}{\tilde{v}_p} - 1\right) \cdot \frac{360°}{\lambda}. \tag{2.30}$$

The phase error is accumulated over propagation distance. Thus, the further the distance the signal propagates across the transmission line, the larger the phase error that is accumulated.

2.5   STABILITY

Recursive time-update procedures, such as the coupled first-order difference operators in (2.13) and (2.14), are subject to a stability criterion. An operator is considered to be stable if the energy in the system remains bounded when the energy input into the system is bounded [3]. The focus of this section is to study the conditions for which the recursive update procedure is stable. The study will be carried out using an eigenvalue analysis of the recursive difference operators. To this end, (2.13) and (2.14) are combined into a discrete-form of the wave equation, as was done in (2.17). The second-order difference operator in (2.17) is here expressed as a linear operator:

$$\mathbf{f}^{n+1} = 2\mathbf{A}\mathbf{f}^n - \mathbf{I}\mathbf{f}^{n-1} \tag{2.31}$$

where, $\mathbf{f}$ is the vector of discrete currents for all i, $\mathbf{I}$ is the identity matrix, and $\mathbf{A}$ is a sparse matrix defined as:

$$\mathbf{A} = \mathbf{I} - \frac{(c\Delta t)^2}{2\Delta x^2}\mathbf{L}. \tag{2.32}$$

$\mathbf{L}$ is a tridiagonal matrix with each row represented by:

$$[\mathbf{L}]_{k-\text{th} \atop \text{row}} = [0, 0, \cdots, 0, -1, 2, -1, 0, \cdots, 0, 0] \tag{2.33}$$

where $L_{k,k} = 2$ is the diagonal term. The matrix $\mathbf{L}$ is positive semi-definite and has eigenvalues $0 \leq \lambda_L \leq 4$.

The stability of the second-order difference operator in (2.31) is simplified by first projecting it as a first-order difference operator with the transformation of variables:

$$\mathbf{y}^{n+1} = \begin{bmatrix} \mathbf{f}^{n+1} \\ \mathbf{f}^n \end{bmatrix}, \quad \text{and } \mathbf{y}^n = \begin{bmatrix} \mathbf{f}^n \\ \mathbf{f}^{n-1} \end{bmatrix}. \tag{2.34}$$

Equation (2.31) can then be expressed as a first-order difference operator:

$$\mathbf{y}^{n+1} = \mathbf{M}\mathbf{y}^n \tag{2.35}$$

where

$$M = \begin{bmatrix} 2A & -I \\ I & 0 \end{bmatrix}. \tag{2.36}$$

Let the initial condition for the problem be specified at $n = 0$. Then, from (2.35), at time-step n,

$$y^{n+1} = M^{n+1}y^0 \tag{2.37}$$

where M^{n+1} is M raised to the $(n+1)$-th power. It can be shown that the total energy on the transmission line is proportional to $|y^{n+1}|^2$, where $|\ \ |$ represents the vector norm. Thus, to guarantee that the solution is bounded, we must ensure that:

$$\lim_{n \to \infty} \frac{|y^n|^2}{|y^0|^2} \leq C \tag{2.38}$$

where C is a constant that is finite. This can only be true if:

$$\lim_{n \to \infty} |M^n| < K \tag{2.39}$$

where $|\ \ |$ represents the matrix norm, and K is a constant that is finite. This will only be satisfied if the spectral radius of M is less than or equal to 1 (i.e., $\rho(M) \leq 1$). This is equivalent to all the eigenvalues of M lie within or on the unit circle in the complex plane [6]. Thus, to predict stability, the eigenspectrum of M must be studied.

Let (λ_A, p) be the eigensolution of the matrix A [6]:

$$Ap = \lambda_A p. \tag{2.40}$$

Also, let $(\lambda_M, [x, y]^T)$ be the eigensolution of the matrix M:

$$M \begin{bmatrix} x \\ y \end{bmatrix} = \lambda_M \begin{bmatrix} x \\ y \end{bmatrix}. \tag{2.41}$$

If the eigenvalues of A are real and distinct, then A is diagonalizable [7]. That is,

$$A = PDP^{-1} \tag{2.42}$$

where P is the matrix of eigenvectors of A, and D is the diagonal matrix of the eigenvalues of A. Combining (2.42) and (2.36), (2.41) is diagonalized:

$$\begin{bmatrix} P & 0 \\ 0 & P \end{bmatrix} \begin{bmatrix} 2D & -I \\ I & 0 \end{bmatrix} \begin{bmatrix} P^{-1} & 0 \\ 0 & P^{-1} \end{bmatrix} \begin{bmatrix} x \\ y \end{bmatrix} = \lambda_M \begin{bmatrix} x \\ y \end{bmatrix}. \tag{2.43}$$

Then, substituting

$$\begin{bmatrix} x' \\ y' \end{bmatrix} = \begin{bmatrix} P^{-1} & 0 \\ 0 & P^{-1} \end{bmatrix} \begin{bmatrix} x \\ y \end{bmatrix} \tag{2.44}$$

Equation (2.43) can be re-cast as:

$$\begin{bmatrix} 2\mathbf{D} & -\mathbf{I} \\ \mathbf{I} & 0 \end{bmatrix} \begin{bmatrix} \mathbf{x}' \\ \mathbf{y}' \end{bmatrix} = \lambda_M \begin{bmatrix} \mathbf{x}' \\ \mathbf{y}' \end{bmatrix}. \tag{2.45}$$

Since the matrix is block diagonal, each eigenvalue ξ can be determined by the simple two-by-two matrix equation:

$$\begin{bmatrix} 2\lambda_A & -1 \\ 1 & 0 \end{bmatrix} \begin{bmatrix} x' \\ y' \end{bmatrix} = \lambda_M \begin{bmatrix} x' \\ y' \end{bmatrix} \tag{2.46}$$

where λ_A is an eigenvalue of $\mathbf{A}$. The eigenvalues of (2.46) can be derived from the characteristic polynomial [6]

$$\lambda_M^2 - 2\lambda_A \lambda_M + 1 = 0, \tag{2.47}$$

which has solutions $\lambda_M = \lambda_A \pm \sqrt{\lambda_A^2 - 1}$. Stability requires that $|\lambda_M| \leq 1$, which is true if

$$|\lambda_A| \leq 1. \tag{2.48}$$

Consequently, stability requires that $\rho(\mathbf{A}) \leq 1$.

The eigenvalues of $\mathbf{A}$ are found from (2.40). Inserting (2.32) into this expression:

$$\left\{ \mathbf{I} - \frac{(c\Delta t)^2}{2\Delta x^2} \mathbf{L} \right\} \mathbf{p} - \lambda_A \mathbf{I} \mathbf{p} = 0. \tag{2.49}$$

This can be rearranged as:

$$\mathbf{L}\mathbf{p} = \lambda_L \mathbf{I}\mathbf{p} \tag{2.50}$$

where

$$\lambda_L = (1 - \lambda_A) \frac{2\Delta x^2}{(c\Delta t)^2}. \tag{2.51}$$

λ_L are the eigenvalues of the tri-diagonal matrix $\mathbf{L}$, which are known to be on the range $0 \leq \lambda_L \leq 4$. From (2.51), we can derive

$$\lambda_A = 1 - \lambda_L \frac{(c\Delta t)^2}{2\Delta x^2}. \tag{2.52}$$

From (2.48), it was determined that stability requires $|\lambda_A| \leq 1$. This must be satisfied for the full range of λ_L. As a result,

$$\left| 1 - 4 \frac{(c\Delta t)^2}{2\Delta x^2} \right| \leq 1. \tag{2.53}$$

This inequality will be true providing that:

$$c\Delta t \leq \Delta x. \tag{2.54}$$

As a result, the time-step must be less than the space step divided by the wave speed of the transmission line. This "stability limit" of the time-step is better known as the Courant-Fredrichs-Lewy

(CFL) stability limit [8]. Within this context, the limit restricts the discretized signal from traveling more than a single spatial cell during a single time-step. If the time-step exceeds the CFL limit, the recursive update equations of (2.13) and (2.14) will lead to an unbounded solution. From (2.37), it is seen that the solution will likely grow exponentially. This is due to the explicit nature of the recursive update scheme, in that attempting to propagate the wave further than a space cell will violate causality.

2.6   SOURCES AND LOADS

Assume a transmission line of length d that is excited by a Thévenin voltage source at $x = 0$ and terminated by a load at $x = d$. The line is discretized into N uniform segments of length $\Delta x = d/N$. The voltages and currents are discretized along the line as illustrated in Fig. 2.1, with $V_0^{n+\frac{1}{2}}$ being the voltage at the source, and $V_N^{n+\frac{1}{2}}$ the voltage at the load. Near the source, the current $I_{\frac{1}{2}}^n$ is discretized a half of a cell into the line. Similarly, near the load, the discrete current is $I_{N-\frac{1}{2}}^n$.

When enforcing source and load conditions, a strict $V - I$ relationship must be enforced. Due to the staggered discretization, both V and I are not known at the same point in space or time. Consequently, clever methods must be employed to properly enforce the $V - I$ relationship without losing stability or accuracy. How this is done for a number of canonical loads and sources is the topic of this section.

Short and Open Circuit Loads

Perhaps the simplest loads to implement are short circuit and open circuit loads, since these can be enforced as Dirichlet boundary conditions. For a short circuit load, the load voltage is simply enforced by constraining the load voltage

$$V_N^{n+\frac{1}{2}}\bigg|_{\substack{\text{short} \\ \text{circuit}}} = 0 \tag{2.55}$$

for all time. An open circuit could be enforced as $I_{N-\frac{1}{2}}^n = 0$. The trouble with this is that the current node is displaced from the actual load location at $x = d$. This introduces error since the line is artificially shortened by $\Delta x/2$. An alternative would be to use image theory. To this end, an image current is posed $\Delta x/2$ beyond the load, and it is constrained such that $I_{N+\frac{1}{2}}^n = -I_{N-\frac{1}{2}}^n$. Combining this with (2.13), leads to the recursive update equation:

$$V_N^{n+\frac{1}{2}} = V_N^{n-\frac{1}{2}} + 2\frac{\Delta t}{C\Delta x}I_{N-\frac{1}{2}}^n \tag{2.56}$$

which is second-order accurate, and it is stable within the CFL limit.

Resistive Loads

Now, consider a resistive load R_L at $x = d$. At the load, the line voltage and current must satisfy Ohm's law: $V(d) = I(d) R_L$. In the discrete space, one could naively enforce $V_N^{n+\frac{1}{2}} = I_{N-\frac{1}{2}}^n R_L$. The problem with this is that since the discrete voltage and current are displaced in both time and space, this leads to a first-order accurate approximation at best. It can also be shown that if $R_L > Z_0$, where $Z_0 = \sqrt{L/C}$ (Ω) is the characteristic impedance of the transmission line, this formulation is unconditionally *unstable*. Thus, this is not an acceptable formulation.

A better approach would be to view the discrete transmission line as a distributed circuit, as illustrated in Fig. 2.3. The node voltages of the distributed line represent the discrete voltages from the finite-difference time-domain approximation. Similarly, the series branch currents represent the discrete currents. The distributed transmission line can be thought of as a cascading of pi-networks [9]. This is done by splitting the shunt capacitance in half with two parallel capacitances, as illustrated in Fig. 2.4.

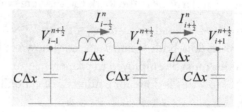

Figure 2.3: Distributed lumped circuit transmission line.

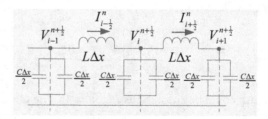

Figure 2.4: Cascade – network representation of the distributed lumped circuit transmission line.

Using the distributed pi-network representation, the transmission line is terminated by a purely resistive load, as illustrated in Fig. 2.5. Kirchoff's Current Law (KCL) is applied at node N, leading to:

$$I_{N-\frac{1}{2}}^n = \frac{C \Delta x}{2} \frac{\partial V_N^{n+\frac{1}{2}}}{\partial t} + \frac{V_N^{n+\frac{1}{2}}}{R_L}. \tag{2.57}$$

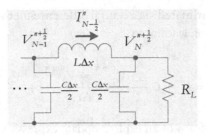

Figure 2.5: Resistive load.

The time-derivative can be expressed as a central difference approximation, which is second-order accurate at time $n\,\Delta t$. Note that the current through the load resistance is expressed at time $(n + \frac{1}{2})\Delta t$, which is not concurrent with the other two branch currents. This can be resolved by applying a linear average in time to the load current. This leads to:

$$I^n_{N-\frac{1}{2}} \approx \frac{C\Delta x}{2} \frac{V^{n+\frac{1}{2}}_N - V^{n-\frac{1}{2}}_N}{\Delta t} + \frac{V^{n+\frac{1}{2}}_N + V^{n-\frac{1}{2}}_N}{2R_L}. \tag{2.58}$$

This can now be used to formulate an explicit update expression at node N for $V^{n+\frac{1}{2}}_N$:

$$V^{n+\frac{1}{2}}_N = \left(\frac{C\Delta x}{2\Delta t} + \frac{1}{2R_L}\right)^{-1} \left(\frac{C\Delta x}{2\Delta t} - \frac{1}{2R_L}\right) V^{n-\frac{1}{2}}_N + \left(\frac{C\Delta x}{2\Delta t} + \frac{1}{2R_L}\right)^{-1} I^n_{N-\frac{1}{2}}. \tag{2.59}$$

Thévenin Voltage Source

Next, consider the excitation by a discrete source with source voltage V_g and internal resistance R_g. This is illustrated in Fig. 2.6, where the source is expressed as a Norton equivalent source. KCL

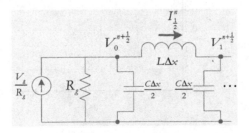

Figure 2.6: Discrete Norton source excitation.

is applied at node 0, leading to:

$$\frac{V^n_g}{R_g} = I^n_{\frac{1}{2}} + C\frac{\Delta x}{2}\frac{\partial V^n_0}{\partial t} + \frac{V^n_0}{R_g}. \tag{2.60}$$

The time derivative is approximated via a central difference method and the voltage across the source resistance is averaged in time, leading to:

$$\frac{V_g^n}{R_g} = I_{\frac{1}{2}}^n + C\frac{\Delta x}{2}\frac{V_0^{n+\frac{1}{2}} - V_0^{n-\frac{1}{2}}}{\Delta t} + \frac{V_0^{n+\frac{1}{2}} + V_0^{n-\frac{1}{2}}}{2R_g}. \tag{2.61}$$

This is rearranged into the form of an explicit update expression at node 0:

$$V_0^{n+\frac{1}{2}} = \left(C\frac{\Delta x}{2\Delta t} + \frac{1}{2R_g}\right)^{-1}\left[\left(C\frac{\Delta x}{2\Delta t} - \frac{1}{2R_g}\right)V_0^{n-\frac{1}{2}} - I_{\frac{1}{2}}^n + \frac{V_g^n}{R_g}\right]. \tag{2.62}$$

Both this expression and (2.59) lead to second-order accuracy, and they are both stable within the CFL limit [9].

As an example, consider a lossless, uniform transmission line with capacitance and inductance per unit length $C = 100$ pF/m and $L = 0.25$ μH/m that is 0.1 m in length. The line has characteristic impedance $Z_0 = \sqrt{L/C} = 50$ Ω. The line is excited by a 1V trapezoidal pulse source with a 50 ohm internal resistance. The trapezoidal pulse has a 40 ps rise time and fall time and a 100 ps duration. The line is also terminated with a matched 50 ohm load. This line is discretized via the FDTD method with 50 uniform segments, or 51 voltage nodes. The time-step is chosen at the CFL limit ($\Delta t = \Delta x/c = \sqrt{LC}\Delta x = 10$ ps). Figure 2.7 illustrates the line voltage at the center node as a function of time as simulated via the FDTD method. A discrete sampling of the exact solution is

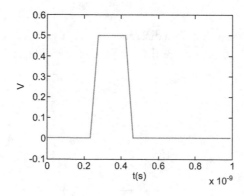

Figure 2.7: Uniform 50 Ω transmission line excited by $1V$ trapezoidal pulse source with 50 Ω source and load resistance.

observed. Note that the amplitude of a trapezoidal pulse is 0.5 V due to the voltage division between the line and the source resistances. Also, note that the pulse is completely absorbed by the matched load.

Next, the source resistance is changed to 100 ohms, and the load resistance is changed to 25 ohms. The line is again simulated. The line voltage recorded at the center node as a function of time is illustrated in Fig. 2.8. The results are exact. Applying voltage division, the initial pulse

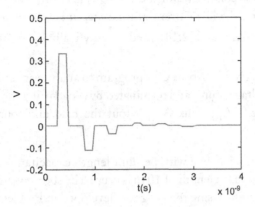

Figure 2.8: Uniform 50 Ω transmission line excited by a $1V$ trapezoidal pulse source with 100 Ω source and 25 Ω load.

amplitude is = 1 V $\cdot$ (50 Ω)/(50 Ω + 100 Ω) = 1/3V. The reflection coefficient due to the load is Γ_L = (25 Ω − 50 Ω)/(25 Ω + 50 Ω) = −1/3. Thus, the reflected pulse has an amplitude of -1/9 V. The source reflection coefficient is Γ_s = (100 Ω − 50 Ω)/ (100 Ω + 50 Ω) = +1/3. Thus, the secondary reflection has an amplitude of -1/27 V. The multiple reflections continue, until a steady-state voltage of 0 volts is realized. This is exactly represented by the results in Fig. 2.8.

2.7   PROBLEMS

1. Starting with (2.13) and (2.14), derive the second-order difference equation for $V_i^{n+\frac{1}{2}}$ that is dual to (2.17).

2. Derive the numerical group velocity $\tilde{v}_g = 1/(d\tilde{k}/d\omega)$ from (2.27). What is the group velocity at the magic time-step? Reproduce the plot in Fig. 2.2 for the group velocity normalized by the wave speed c.

3. Derive (2.31) from (2.17). Then, derive the first-order difference operator in (2.35), with **M** defined by (2.36), from (2.31).

4. Show that (2.38) is proportional to the ratio of the total energy (electric plus magnetic) on the transmission line at time-step n to that at time-step 0. If the transmission line is passive linear, explain why (2.38) must be true in terms of conservation of energy.

5. A uniform transmission-line with per unit length capacitance and inductance C = 100 pF/m and L = 250 nH/m has a length of 1 m and is terminated by a short circuit at each end. A staggered grid similar to Fig. 2.1 with $\Delta x = 0.1$ m is used to represent line voltages and currents. The initial conditions for the line voltage at time-step $n - \frac{1}{2}$ is 1V for $3 \le i \le 6$, and 0 V elsewhere, and the line current at time-step n is 0.02 A for $3.5 \le i + \frac{1}{2} \le 6.5$, and 0 A elsewhere. Using (2.13) and (2.14), predict analytically $V_i^{n+1/2}$ and $I_{i+\frac{1}{2}}^{n+1}$ for all i if $c\Delta t = \Delta x$.

6. Write a FORTRAN, Matlab, or C++ program to analyze an arbitrary transmission line excited by a Thévenin voltage source and terminated by a load using the finite-difference time-domain approximation (c.f., Appendix A). Output the load and source voltages and currents as a function of time.

7. A uniform transmission-line with per unit length capacitance and inductance C = 100 pF/m and L = 250 nH/m has a length of 10 m is excited by source voltage $v(t) = 2u(t)$, where $u(t)$ is the unit-step function. Using the program developed in Problem 5, predict the load and source voltages as a function of time if: (a) $R_g = 50\ \Omega$, $R_L = 50\ \Omega$, (b) $R_g = 25\ \Omega$, $R_L = 100\ \Omega$, and (c) $R_g = 100\ \Omega$, $R_L = 20\ \Omega$. Choose $N = 50$, and use the magic time-step.

8. Repeat Problem 6 using $\Delta t = 0.99\Delta x/c$. Explain your results.

9. Following the procedure used for resistive loads in Section 2.6, derive an explicit update expression for a parallel capacitor-resistor load.

10. Repeat Problem 8 for an series inductor-resistor load. [Hint: A load current will have to be introduced, and both KVL (Kirchoff's voltage law) and KCL must be enforced about the load.]

11. Repeat Problem 9 for a parallel RLC load.

REFERENCES

[1] C. R. Paul, *Analysis of Multiconductor Transmission Lines*, 2nd ed. Piscataway, NJ: Wiley-IEEE Press, 2007. DOI: 10.1109/9780470547212 21

[2] G. B. Thomas, Jr. and R. L. Finney, *Calculus and Analytical Geometry*, 9th ed. Reading, MA: Addison-Wesley, 1996. 22, 26

[3] F. B. Hildebrand, *Finite-Difference Equations and Simulations*. Englewood Cliffs, New Jersey: Pretice-Hall, 1968. 22, 28

[4] R. F. Harrington, *Time-Harmonic Electromagnetic Fields*. New York: McGraw-Hill, 1961. 26, 27

[5] A. Taflove, "Computational Electrodynamics: The Finite Difference Time Domain," Boston: Artech House, 1995. 27

[6] C. Lanzos, *Applied Analysis*. Englewood Cliffs, NJ: Prentice-Hall, Inc., 1964. 29, 30

[7] G. H. Golub and C. F. Van Loan, *Matrix Computations*, 2nd ed. Baltimore: The Johns Hopkins University Press, 1989. 29

[8] R. Courant, K. O. Friedrichs, and H. Lewy, "On the Partial Differential Equations of Mathematical Physics (translated)," *IBM Journal of Research and Development*, vol. 11, pp. 215–234, 1967. DOI: 10.1147/rd.112.0215 31

[9] A. Orlandi and C. R. Paul, "FDTD analysis of lossy, multiconductor transmission lines terminated in arbitrary loads," *IEEE Transactions on Electromagnetic Compatibility*, vol. 38, pp. 388–399, Aug 1996. DOI: 10.1109/15.536069 32, 34

CHAPTER 3

Yee Algorithm for Maxwell's Equations

3.1 MAXWELL'S EQUATIONS

The mathematical relationship of the electromagnetic fields radiated by time-dependent current or charge densities is governed by Maxwell's equations. These can be expressed in differential form as:

$$\text{Faraday's Law:} \quad \frac{\partial \vec{B}}{\partial t} = -\nabla \times \vec{E} - \vec{M} \tag{3.1}$$

$$\text{Ampere's Law:} \quad \frac{\partial \vec{D}}{\partial t} = \nabla \times \vec{H} - \vec{J} \tag{3.2}$$

$$\text{Gauss's Laws:} \quad \nabla \cdot \vec{D} = \rho$$
$$\nabla \cdot \vec{B} = \rho^* \tag{3.3}$$

$$\text{Continuity Equations:} \quad \nabla \cdot \vec{J} = -\frac{\partial}{\partial t}\rho$$
$$\nabla \cdot \vec{M} = -\frac{\partial}{\partial t}\rho^* \tag{3.4}$$

where, using MKS units, $\vec{B}$ is the magnetic flux density (Wb/m^2), $\vec{D}$ is the electric flux density (C/m^2), $\vec{E}$ is the electric field intensity (V/m), $\vec{H}$ is the magnetic field intensity (A/m), $\vec{J}$ is the electric current density (A/m^2), $\vec{M}$ is the magnetic current density (V/m^2), ρ is the electric charge density (C/m^3), and ρ^* is the magnetic charge density (Wb/m^3). The flux densities and the field intensities are related through the constitutive relations. For linear, isotropic media, these are:

$$\vec{D} = \varepsilon \vec{E} = \varepsilon_o \varepsilon_r \vec{E} \tag{3.5}$$

$$\vec{B} = \mu \vec{H} = \mu_o \mu_r \vec{H} \tag{3.6}$$

where ε_o is the free-space permittivity (8.854×10^{-12} F/m), ε_r is the relative permittivity, and ε is the permittivity (F / m) of the media. Similarly, μ_o is the free-space permeability ($4\pi \times 10^{-7}$ H/m), μ_r is the relative permeability, and μ is the permeability (H / m) of the media.

Maxwell's equations can also be cast into an integral form:

Faraday's Law: $\quad \dfrac{\partial}{\partial t} \iint_S \vec{B} \cdot d\vec{S} = -\oint_C \vec{E} \cdot d\vec{\ell} - \iint_S \vec{M} \cdot d\vec{s}$ (3.7)

Ampere's Law: $\quad \dfrac{\partial}{\partial t} \iint_S \vec{D} \cdot d\vec{S} = \oint_C \vec{H} \cdot d\vec{\ell} - \iint_S \vec{J} \cdot d\vec{s}$ (3.8)

Gauss's Laws: $\quad \oiint_S \vec{D} \cdot d\vec{s} = \iiint_V \rho \, dv$

$$\oiint_S \vec{B} \cdot d\vec{s} = \iiint_V \rho^* \, dv \qquad (3.9)$$

Continuity Equations: $\quad \oiint_S \vec{J} \cdot d\vec{s} = -\dfrac{\partial}{\partial t} \iiint_V \rho \, dv$

$$\oiint_S \vec{M} \cdot d\vec{s} = -\dfrac{\partial}{\partial t} \iiint_V \rho^* \, dv \qquad (3.10)$$

where in (3.7) and (3.8), S is an open surface bound by the contour C, and in (3.9) and (3.10), V is a volume bound by the closed surface S.

The electromagnetic fields and flux densities are subject to boundary conditions. These boundary conditions can be derived directly from Maxwell's equations in their integral form. Assume a current and charge-free boundary S that separates two regions, V_1 and V_2, with material profiles (ε_1, μ_1) and (ε_2, μ_2), respectively. Faraday's and Ampère's laws predict the tangential electric and magnetic fields are continuous across the boundary. This is expressed via the tangential boundary conditions

$$\hat{n} \times \left(\vec{E}_1 - \vec{E}_2 \right)\Big|_S = 0, \qquad \hat{n} \times \left(\vec{H}_1 - \vec{H}_2 \right)\Big|_S = 0 \qquad (3.11)$$

where $\hat{n}$ is the unit vector normal to S directed into V_1. Gauss's laws predict that the normal electric and magnetic fluxes are continuous across the boundary. This is expressed via the normal boundary conditions

$$\hat{n} \cdot \left(\vec{D}_1 - \vec{D}_2 \right)\Big|_S = 0, \qquad \hat{n} \cdot \left(\vec{B}_1 - \vec{B}_2 \right)\Big|_S = 0. \qquad (3.12)$$

If the surface S supports surface current densities $\vec{J}_s$, $\vec{M}_s$ and/or surface charge densities ρ_s, ρ_s^*, then Faraday's and Ampère's laws and Gauss's laws predict the boundary conditions:

$$\hat{n} \times \left(\vec{E}_1 - \vec{E}_2 \right)\Big|_S = -\vec{M}_s, \qquad \hat{n} \times \left(\vec{H}_1 - \vec{H}_2 \right)\Big|_S = \vec{J}_s \qquad (3.13)$$

$$\hat{n} \cdot \left(\vec{D}_1 - \vec{D}_2 \right)\Big|_S = \rho_s, \qquad \hat{n} \cdot \left(\vec{B}_1 - \vec{B}_2 \right)\Big|_S = \rho_s^*. \qquad (3.14)$$

There are four boundary conditions derived from Maxwell's equations. However, in general, only two need to be applied for uniqueness when posing solutions for the fields and fluxes. Note that the pairing is not arbitrary. One can either pair the continuity of the two tangential field vectors, or, one can pair the continuity of a tangential field, with the corresponding normal flux density (i.e., $\vec{E}$ and $\vec{D}$). Interestingly, it can be shown that if tangential $\vec{E}$ is continuous, then by nature of the Maxwell's equations, this also constrains normal $\vec{B}$ to be continuous. This is similar for $\vec{H}$ and $\vec{D}$. Consequently, these boundary conditions would not represent an independent pairing needed for a unique solution.

Special boundary conditions also need to be posed for perfectly conducting media. For example, consider the case where V_2 is a perfect electrical conductor (PEC) bounded by surface S_{PEC}. By definition, the electric and magnetic fields within the PEC are 0 (i.e., $\vec{E}_2 = \vec{H}_2 = 0$). Consequently, PEC surface boundary conditions are:

$$\hat{n} \times \vec{E}_1 \Big|_{S_{PEC}} = 0, \qquad \hat{n} \times \vec{H}_1 \Big|_{S_{PEC}} = \vec{J}_s \qquad (3.15)$$

$$\hat{n} \cdot \vec{D}_1 \Big|_{S_{PEC}} = \rho_s, \qquad \hat{n} \cdot \vec{B}_1 \Big|_{S_{PEC}} = 0 \qquad (3.16)$$

where $\vec{J}_s$ is the surface electric current density *induced* on S_{PEC}, and ρ_s is the electric charge density induced on S_{PEC}. From the continuity equation, $\nabla_s \cdot \vec{J}_s = -\partial/\partial t \rho_s$. By duality, on the surface of a perfect magnetic conductor (PMC) boundary, the fields and fluxes satisfy the PMC surface boundary conditions:

$$\hat{n} \times \vec{E}_1 \Big|_{S_{PMC}} = -\vec{M}_s, \qquad \hat{n} \times \vec{H}_1 \Big|_{S_{PMC}} = 0 \qquad (3.17)$$

$$\hat{n} \cdot \vec{D}_1 \Big|_{S_{PMC}} = 0, \qquad \hat{n} \cdot \vec{B}_1 \Big|_{S_{PMC}} = \rho_s^*. \qquad (3.18)$$

A more comprehensive description of Maxwell's equations and analytical solutions of these equations can be found in classical references such as [1, 2].

3.2 THE YEE-ALGORITHM

In 1966, Kane S. Yee derived an elegant, yet simple, time-dependent solution of Maxwell's equations based on their differential form using central difference approximations of both the space and the time-derivatives [3]. The formulation is based on discretizing the volume domain with a regular, structured, staggered, rectangular grid. Similar to the one-dimensional transmission-line equations, Yee discovered that in order to maintain second-order accuracy of the central difference operators, the electric and magnetic fields must be staggered in both space and time. The novel scheme he derived to achieve this, now referred to as the Yee-algorithm, is detailed in this section.

Consider a uniformly spaced rectangular grid in three-dimensions. Each grid cell has dimensions Δx, Δy, and Δz along each Cartesian axis. The coordinate of a node of the grid can be expressed in discrete form as: $(x, y, z)_{i,j,k} = (i\Delta x, j\Delta y, k\Delta z)$, where i, j, and k are integers. Similarly, time is uniformly discretized as $t = n\Delta t$. An arbitrary function $f(x, y, z, t)$ can be expressed at any node within the discrete space using the notation:

$$f(x, y, z, t) = f(i\Delta x, j\Delta y, k\Delta z, n\Delta t) = f_{i,j,k}^n. \tag{3.19}$$

Within this uniform grid, the projections of the vector electric field parallel to a grid edges are sampled at edge grid edge center. Dual to this, the projection of the magnetic field normal to each grid cell face is sampled at the center of a grid face. This is illustrated in Fig. 3.1. Observing Fig. 3.1, it is apparent that the tangential electric field projected on the edges bounding a cell face circulate about the normal magnetic field vectors. This provides the essential pieces to formulate a

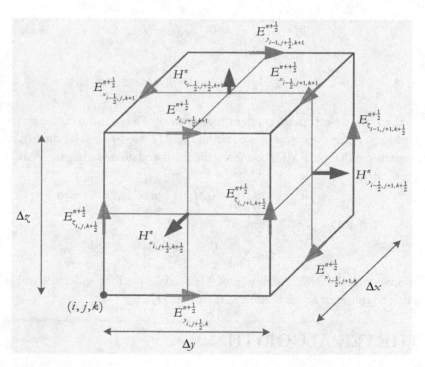

Figure 3.1: Primary grid cell of the regular, structured, rectangular, staggered grip.

curl operation. To illustrate this, consider the x-projection of Faraday's law in (3.1):

$$\mu \frac{\partial H_x}{\partial t} = \frac{\partial E_y}{\partial z} - \frac{\partial E_z}{\partial y} - M_x \tag{3.20}$$

where a linear isotropic material is assumed. Using the discretization of Fig. 3.1, the time derivative and the spatial derivatives from the curl operator are approximated via central differences, leading to a discrete form of Faraday's law:

$$
\mu \left(\frac{H_x^{n+1}_{\,i,j+\frac{1}{2},k+\frac{1}{2}} - H_x^{n}_{\,i,j+\frac{1}{2},k+\frac{1}{2}}}{\Delta t} \right) = \left(\frac{E_y^{n+\frac{1}{2}}_{\,i,j+\frac{1}{2},k+1} - E_y^{n+\frac{1}{2}}_{\,i,j+\frac{1}{2},k}}{\Delta z} \right)
$$

$$
- \left(\frac{E_z^{n+\frac{1}{2}}_{\,i,j+1,k+\frac{1}{2}} - E_z^{n+\frac{1}{2}}_{\,i,j,k+\frac{1}{2}}}{\Delta y} \right) - M_x^{n+\frac{1}{2}}_{\,i,j+\frac{1}{2},k+\frac{1}{2}}. \tag{3.21}
$$

It is observed from (3.21) that the central difference approximations of both spatial derivatives of the electric field projections are second order accurate at the face center, which is the sample location of the normal magnetic field. By staggering the magnetic field and the electric field in time, the time-derivative is also second-order accurate. Consequently, the difference operator in (3.21) is second-order accurate in both space and time. Following the same procedure, one can derive similar expressions for the y and z-projections of Faraday's law as well:

$$
\mu \left(\frac{H_y^{n+1}_{\,i+\frac{1}{2},j,k+\frac{1}{2}} - H_y^{n}_{\,i+\frac{1}{2},j,k+\frac{1}{2}}}{\Delta t} \right) =
$$
$$
\left(\frac{E_z^{n+\frac{1}{2}}_{\,i+1,j,k+\frac{1}{2}} - E_z^{n+\frac{1}{2}}_{\,i,j,k+\frac{1}{2}}}{\Delta x} \right) - \left(\frac{E_x^{n+\frac{1}{2}}_{\,i+\frac{1}{2},j,k+1} - E_x^{n+\frac{1}{2}}_{\,i+\frac{1}{2},j,k}}{\Delta z} \right) - M_y^{n+\frac{1}{2}}_{\,i+\frac{1}{2},j,k+\frac{1}{2}}, \tag{3.22}
$$

$$
\mu \left(\frac{H_z^{n+1}_{\,i+\frac{1}{2},j+\frac{1}{2},k} - H_z^{n}_{\,i+\frac{1}{2},j+\frac{1}{2},k}}{\Delta t} \right) =
$$
$$
\left(\frac{E_x^{n+\frac{1}{2}}_{\,i+\frac{1}{2},j+1,k} - E_x^{n+\frac{1}{2}}_{\,i+\frac{1}{2},j,k}}{\Delta y} \right) - \left(\frac{E_y^{n+\frac{1}{2}}_{\,i+1,j+\frac{1}{2},k} - E_y^{n+\frac{1}{2}}_{\,i,j+\frac{1}{2},k}}{\Delta x} \right) - M_z^{n+\frac{1}{2}}_{\,i+\frac{1}{2},j+\frac{1}{2},k}. \tag{3.23}
$$

The discrete form of Ampère's law is derived via a secondary grid cell, as illustrated in Fig. 3.2. The secondary grid cell edges connect the cell centers of the primary grid cells illustrated in Fig. 3.1. The secondary grid cell also has dimensions Δx, Δy, and Δz. Thus, the edges of the secondary grid pass through the centers of the faces of the primary grid cells. Dually, the edges of the primary grid

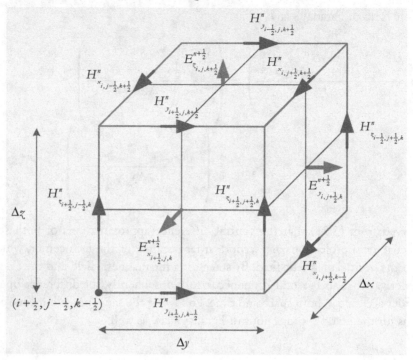

Figure 3.2: Secondary grid cell of the regular, structured, rectangular, staggered grip.

pass through the face centers of the secondary grid cells. Consequently, the electric and magnetic field vectors have dual roles in the primary and secondary grids.

Observing a secondary grid face in Fig. 3.2, it is apparent that the magnetic field lines on a cell face circulates about the normal electric field line. Again, this provides the essential components of a curl operation. The x-projection of Ampère's law is expressed in a discrete form as:

$$\varepsilon \left(\frac{E_{x}^{n+\frac{1}{2}}_{i+\frac{1}{2},j,k} - E_{x}^{n-\frac{1}{2}}_{i+\frac{1}{2},j,k}}{\Delta t} \right) = \left(\frac{H_{z}^{n}_{i+\frac{1}{2},j+\frac{1}{2},k} - H_{z}^{n}_{i+\frac{1}{2},j-\frac{1}{2},k}}{\Delta y} \right)$$

$$- \left(\frac{H_{y}^{n}_{i+\frac{1}{2},j,k+\frac{1}{2}} - H_{y}^{n}_{i+\frac{1}{2},j,k-\frac{1}{2}}}{\Delta z} \right) - J_{x}^{n}_{i+\frac{1}{2},j,k} \qquad (3.24)$$

where an isotropic, linear, lossless media has been assumed. Similar expressions can be derived for the y and z-projections of Ampère's law.

The discrete form of Faraday's and Ampère's laws lead to a total of six equations, which can then be used to solve for the time-dependent vector field intensities. Yee proposed to do this with an explicit time-marching scheme. To this end, it is assumed that the initial values of the discrete fields are known over all space. Subsequently, a recursive solution scheme can be used to advance the fields through time. For example, from (3.21), assuming that $E_y^{n+\frac{1}{2}}$, $E_z^{n+\frac{1}{2}}$, and H_x^n are known at all spatial samples, an explicit update operator used to solve for H_x^{n+1} is expressed as:

$$
H_x^{n+1}{}_{i,j+\frac{1}{2},k+\frac{1}{2}} =
$$

$$
H_x^n{}_{i,j+\frac{1}{2},k+\frac{1}{2}} + \frac{\Delta t}{\mu} \left[\left(\frac{E_y^{n+\frac{1}{2}}{}_{i,j+\frac{1}{2},k+1} - E_y^{n+\frac{1}{2}}{}_{i,j+\frac{1}{2},k}}{\Delta z} \right) - \left(\frac{E_z^{n+\frac{1}{2}}{}_{i,j+1,k+\frac{1}{2}} - E_z^{n+\frac{1}{2}}{}_{i,j,k+\frac{1}{2}}}{\Delta y} \right) - M_x^{n\,|\,\frac{1}{2}}{}_{i,j+\frac{1}{2},k+\frac{1}{2}} \right] \cdot
$$

(3.25)

Similarly,

$$
H_y^{n+1}{}_{i+\frac{1}{2},j,k+\frac{1}{2}} =
$$

$$
H_y^n{}_{i+\frac{1}{2},j,k+\frac{1}{2}} + \frac{\Delta t}{\mu} \left[\left(\frac{E_z^{n+\frac{1}{2}}{}_{i+1,j,k+\frac{1}{2}} - E_z^{n+\frac{1}{2}}{}_{i,j,k+\frac{1}{2}}}{\Delta x} \right) - \left(\frac{E_x^{n+\frac{1}{2}}{}_{i+\frac{1}{2},j,k+1} - E_x^{n+\frac{1}{2}}{}_{i+\frac{1}{2},j,k}}{\Delta z} \right) - M_y^{n+\frac{1}{2}}{}_{i+\frac{1}{2},j,k+\frac{1}{2}} \right]
$$

(3.26)

$$
H_z^{n+1}{}_{i+\frac{1}{2},j+\frac{1}{2},k} =
$$

$$
H_z^n{}_{i+\frac{1}{2},j+\frac{1}{2},k} + \frac{\Delta t}{\mu} \left[\left(\frac{E_x^{n+\frac{1}{2}}{}_{i+\frac{1}{2},j+1,k} - E_x^{n+\frac{1}{2}}{}_{i+\frac{1}{2},j,k}}{\Delta y} \right) - \left(\frac{E_y^{n+\frac{1}{2}}{}_{i+1,j+\frac{1}{2},k} - E_y^{n+\frac{1}{2}}{}_{i,j+\frac{1}{2},k}}{\Delta x} \right) - M_z^{n+\frac{1}{2}}{}_{i+\frac{1}{2},j+\frac{1}{2},k} \right]
$$

(3.27)

$$
E_x^{n+\frac{1}{2}}{}_{i+\frac{1}{2},j,k} =
$$

$$
E_x^{n-\frac{1}{2}}{}_{i+\frac{1}{2},j,k} + \frac{\Delta t}{\varepsilon} \left[\left(\frac{H_z^n{}_{i+\frac{1}{2},j+\frac{1}{2},k} - H_z^n{}_{i+\frac{1}{2},j-\frac{1}{2},k}}{\Delta y} \right) - \left(\frac{H_y^n{}_{i+\frac{1}{2},j,k+\frac{1}{2}} - H_y^n{}_{i+\frac{1}{2},j,k-\frac{1}{2}}}{\Delta z} \right) - J_x^n{}_{i+\frac{1}{2},j,k} \right]
$$

(3.28)

$$E_{y_{i,j+\frac{1}{2},k}}^{n+\frac{1}{2}} =$$

$$E_{y_{i,j+\frac{1}{2},k}}^{n-\frac{1}{2}} + \frac{\Delta t}{\varepsilon} \left[\left(\frac{H_{x_{i,j+\frac{1}{2},k+\frac{1}{2}}}^{n} - H_{x_{i,j+\frac{1}{2},k-\frac{1}{2}}}^{n}}{\Delta z} \right) - \left(\frac{H_{z_{i+\frac{1}{2},j+\frac{1}{2},k}}^{n} - H_{z_{i-\frac{1}{2},j+\frac{1}{2},k}}^{n}}{\Delta x} \right) - J_{y_{i,j+\frac{1}{2},k}}^{n} \right]$$

$$(3.29)$$

$$E_{z_{i,j,k+\frac{1}{2}}}^{n+\frac{1}{2}} =$$

$$E_{z_{i,j,k+\frac{1}{2}}}^{n-\frac{1}{2}} + \frac{\Delta t}{\varepsilon} \left[\left(\frac{H_{y_{i+\frac{1}{2},j,k+\frac{1}{2}}}^{n} - H_{y_{i-\frac{1}{2},j,k+\frac{1}{2}}}^{n}}{\Delta x} \right) - \left(\frac{H_{x_{i,j+\frac{1}{2},k+\frac{1}{2}}}^{n} - H_{x_{i,j-\frac{1}{2},k+\frac{1}{2}}}^{n}}{\Delta y} \right) - J_{z_{i,j,k+\frac{1}{2}}}^{n} \right].$$

$$(3.30)$$

Equations (3.25)–(3.30) are the first-order difference equations defining Yee's algorithm and are the foundation of the FDTD method [4]. These equations provide an explicit recursive update scheme of the electromagnetic fields in linear, isotropic, lossless media throughout the entire volume. The time-step is limited by a stability criterion, which will be discussed in Section 3.5. If the time-step is bound by the stability limit, then this recursive update scheme is second-order accurate in both space and time.

3.3   GAUSS'S LAWS

The discrete approximations of Maxwell's curl equations must also satisfy Gauss's laws. If they do not, then spurious charge can corrupt the numerical solution. First consider Gauss's law for the magnetic field as presented in (3.3). Assuming a charge-free region, then:

$$\nabla \cdot \vec{B} = 0 \tag{3.31}$$

must hold true. If this is differentiated with respect to time, then:

$$\frac{\partial}{\partial t} \nabla \cdot \vec{B} = 0 \tag{3.32}$$

must also be true. The spatial derivative is then approximated with a central difference approximation, leading to a discrete form of Gauss's law:

$$\frac{\partial}{\partial t}\left(\frac{B_x^{n+1}_{i,j+\frac{1}{2},k+\frac{1}{2}} - B_x^{n+1}_{i-1,j+\frac{1}{2},k+\frac{1}{2}}}{\Delta x} + \frac{B_y^{n+1}_{i+\frac{1}{2},j,k+\frac{1}{2}} - B_y^{n+1}_{i+\frac{1}{2},j-1,k+\frac{1}{2}}}{\Delta y}\right.$$

$$\left. + \frac{B_z^{n+1}_{i+\frac{1}{2},j+\frac{1}{2},k} - B_z^{n+1}_{i+\frac{1}{2},j+\frac{1}{2},k-1}}{\Delta z}\right) = 0. \tag{3.33}$$

Substituting in the discrete expressions for the right-hand-sides of (3.21)–(3.23) (with the M's = 0), then it is found that the discrete electric fields identically cancel, and (3.33) is exactly satisfied! A dual expression can be derived for Gauss's law for the electric field.

Thus, it is concluded that the discrete representation satisfies Gauss's law to the extent that the total charge density in the discrete system is constant. Therefore, if the initial charge density is zero, then the total charge density in the discrete system will remain zero, and Gauss's law is strictly satisfied.

3.4   FINITE INTEGRATION TECHNIQUE [5]

The Yee-algorithm was derived directly from the differential form of Maxwell's equations. Further insight to the discretization can be gained by deriving a similar set of equations based on the integral form of Maxwell' equations based on the same staggered grid representation of the fields as illustrated in Figs. 3.1 and 3.2. Consider a face of the primary grid with an x-normal (c.f., Fig. 3.1). Assuming a linear, isotropic, lossless media, Faraday's law in (3.7) expressed over the face is written as:

$$\frac{\partial}{\partial t}\mu \iint_S \vec{H} \cdot \hat{x}dydz = -\oint_C \vec{E} \cdot d\vec{\ell} - \iint_S \vec{M} \cdot \hat{x}dydz \tag{3.34}$$

where the contour integral is performed over the bounding edges. At this point, the following discrete approximation is made: $i)$ the electric field has a constant tangential projection along the length of each primary grid edge, and $ii)$ the magnetic field has a constant normal projection over the entire primary grid face. With this discrete representation, (3.34) is approximated as:

$$\frac{\partial}{\partial t}\mu H_x^{n+\frac{1}{2}}_{i,j+\frac{1}{2},k+\frac{1}{2}}\Delta y\Delta z =$$

$$-(E_y^{n+\frac{1}{2}}_{i,j+\frac{1}{2},k}\Delta y + E_z^{n+\frac{1}{2}}_{i,j+1,k+\frac{1}{2}}\Delta z - E_y^{n+\frac{1}{2}}_{i,j+\frac{1}{2},k+1}\Delta y - E_z^{n+\frac{1}{2}}_{i,j,k+\frac{1}{2}}\Delta z) - M_x^{n+\frac{1}{2}}_{i,j+\frac{1}{2},k+\frac{1}{2}}\Delta y\Delta z \tag{3.35}$$

A central difference approximation is used to evaluate the time derivative, leading to:

$$\mu \frac{H_x^{n+1}\Big|_{i,j+\frac{1}{2},k+\frac{1}{2}} - H_x^n\Big|_{i,j+\frac{1}{2},k+\frac{1}{2}}}{\Delta t} \Delta y \Delta z =$$

$$-(E_y^{n+\frac{1}{2}}\Big|_{i,j+\frac{1}{2},k} \Delta y + E_z^{n+\frac{1}{2}}\Big|_{i,j+1,k+\frac{1}{2}} \Delta z - E_y^{n+\frac{1}{2}}\Big|_{i,j+\frac{1}{2},k+1} \Delta y - E_z^{n+\frac{1}{2}}\Big|_{i,j,k+\frac{1}{2}} \Delta z) - M_x^{n+\frac{1}{2}}\Big|_{i,j+\frac{1}{2},k+\frac{1}{2}} \Delta y \Delta z$$

$$. \quad (3.36)$$

Both sides are multiplied by $1/\Delta y \Delta z$, leading to:

$$\mu \frac{H_x^{n+1}\Big|_{i,j+\frac{1}{2},k+\frac{1}{2}} - H_x^n\Big|_{i,j+\frac{1}{2},k+\frac{1}{2}}}{\Delta t} = \left(\frac{E_y^{n+\frac{1}{2}}\Big|_{i,j+\frac{1}{2},k+1} - E_y^{n+\frac{1}{2}}\Big|_{i,j+\frac{1}{2},k}}{\Delta z} \right)$$

$$- \left(\frac{E_z^{n+\frac{1}{2}}\Big|_{i,j+1,k+\frac{1}{2}} - E_y^{n+\frac{1}{2}}\Big|_{i,j,k+\frac{1}{2}}}{\Delta y} \right) - M_x^{n+\frac{1}{2}}\Big|_{i,j+\frac{1}{2},k+\frac{1}{2}} \quad (3.37)$$

which is identical to (3.21). Consequently, the finite-integration method applied to a uniform rectangular grid leads to an identical difference operator as that derived via the central difference approximation.

It can be shown that the difference operators representing Ampère's law can be derived via a dual discretization based on the secondary grid face. That is, the magnetic field has a constant tangential projection along the length of each secondary grid edge, and the electric field has a constant normal projection over the entire secondary grid face. (c.f., Fig. 3.2). Following this discretization, the difference operators in (3.28)–(3.30) can be derived directly from Ampère's law in its integral form.

Finally, it can also be shown that Gauss's law in the differential form is satisfied in an identical manner as the differential form discussed in Section 3.3.

3.5   STABILITY

In Section 2.5, it was shown that the explicit time-updating procedure of the one-dimensional transmission-line equations was conditionally stable. That is, for linear materials, one expects that the total electric and magnetic energies in the system must remain bounded for all time-steps. It was found that if the eigenvalues of the discrete linear operator moved out of the unit circle, then the system was unstable, as this condition would lead to unbounded energy growth. A similar condition holds true for the three-dimensional Maxwell's equations. As a result, a stability analysis must be performed in order to determine the Courant-Fredrichs-Lewy (CFL) stability limit [6].

The recursive update equations (3.25)–(3.30) spanning the entire problem domain can be expressed as a linear system of equations as:

$$\mathbf{h}^n = \mathbf{h}^{n-1} - \Delta t \mu^{-1} \mathbf{C}_e \mathbf{e}^{n-\frac{1}{2}} \tag{3.38}$$

$$\mathbf{e}^{n+\frac{1}{2}} = \mathbf{e}^{n-\frac{1}{2}} + \Delta t \varepsilon^{-1} \mathbf{C}_h \mathbf{h}^n \tag{3.39}$$

where $\mathbf{h}$ is the vector of all discrete magnetic fields, and $\mathbf{e}$ is the vector of all the discrete electric fields, $\mathbf{C}_e$ is a sparse matrix representing the discrete curl of the electric field, and $\mathbf{C}_h$ is a sparse matrix representing the discrete curl of the magnetic field. Also, for simplicity, the medium is assumed to be infinite and homogeneous.

In order to study the stability of the coupled linear system of equations, it is convenient to cast it into a form of a single first-order difference equation. To arrive at this, we first substitute (3.38) into (3.39), leading to:

$$\mathbf{e}^{n+\frac{1}{2}} = \Delta t \varepsilon^{-1} \mathbf{C}_h \mathbf{h}^{n-1} + \mathbf{e}^{n-\frac{1}{2}} - \Delta t^2 \mu^{-1} \varepsilon^{-1} \mathbf{C}_e \mathbf{C}_h \mathbf{e}^{n-\frac{1}{2}}. \tag{3.40}$$

This is then combined with (3.38), leading to the combined linear system of equations:

$$\begin{pmatrix} \mathbf{h}^n \\ \mathbf{e}^{n+\frac{1}{2}} \end{pmatrix} = \begin{pmatrix} \mathbf{I} & -\Delta t \mu^{-1} \mathbf{C}_e \\ \Delta t \varepsilon^{-1} \mathbf{C}_h & \mathbf{I} - \Delta t^2 \mu^{-1} \varepsilon^{-1} \mathbf{C}_e \mathbf{C}_h \end{pmatrix} \begin{pmatrix} \mathbf{h}^{n-1} \\ \mathbf{e}^{n-\frac{1}{2}} \end{pmatrix} \tag{3.41}$$

where $\mathbf{I}$ is the identity-matrix. This can be expressed in a reduced form as [7]:

$$\mathbf{w}^n = \mathbf{G} \mathbf{w}^{n-1} \tag{3.42}$$

where,

$$\mathbf{w}^n = \begin{pmatrix} \mathbf{h}^n \\ \mathbf{e}^{n+\frac{1}{2}} \end{pmatrix}, \quad \text{and} \quad \mathbf{G} = \begin{pmatrix} \mathbf{I} & -\Delta t \mu^{-1} \mathbf{C}_e \\ \Delta t \varepsilon^{-1} \mathbf{C}_h & \mathbf{I} - c^2 \Delta t^2 \mathbf{C}_e \mathbf{C}_h \end{pmatrix} \tag{3.43}$$

and $c^2 = \mu^{-1} \varepsilon^{-1}$, where c is the speed of light in the material medium.

Let $\mathbf{w}^0$ represent the initial condition or the input to the system. Then, from (3.42), the discrete fields at time-step n can be expressed as:

$$\mathbf{w}^n = \mathbf{G}^n \mathbf{w}^0 \tag{3.44}$$

where $\mathbf{G}^n$ represents $\mathbf{G}$ raised to the nth power. To ensure that the passive linear system is stable, $\lim_{n \to \infty} ||\mathbf{w}^n|| / ||\mathbf{w}^0|| \leq C$, where C is a finite constant. A necessary condition for this to be true is that the eigenvalues of $\mathbf{G}$ lie within or on the unit circle of the complex plane. That is, $|\lambda_G| \leq 1$, where λ_G are the eigenvalues of $\mathbf{G}$.

The eigenvalues of $\mathbf{G}$ are found from the eigenvalue equation [8]:

$$\mathbf{G}\mathbf{x} = \lambda_G \mathbf{x} \tag{3.45}$$

where $\mathbf{x}$ represents the eigenvectors of $\mathbf{G}$. The eigenvalues are shifted via the transformation:

$$(\mathbf{G} - \mathbf{I})\mathbf{x} = (\lambda_G - 1)\mathbf{x} \rightarrow \tilde{\mathbf{G}}\mathbf{x} = \xi\mathbf{x} \tag{3.46}$$

where $\xi = \lambda_G - 1$. This is written more explicitly as:

$$\begin{pmatrix} 0 & -\Delta t \mu^{-1}\mathbf{C}_e \\ \Delta t \varepsilon^{-1}\mathbf{C}_h & -c^2 \Delta t^2 \mathbf{C}_e \mathbf{C}_h \end{pmatrix} \begin{pmatrix} \mathbf{x}_1 \\ \mathbf{x}_2 \end{pmatrix} = \xi \begin{pmatrix} \mathbf{x}_1 \\ \mathbf{x}_2 \end{pmatrix}. \tag{3.47}$$

The first row of (3.47) is multiplied by $\Delta t \varepsilon^{-1}\mathbf{C}_h$, leading to:

$$-\Delta t^2 \mu^{-1}\varepsilon^{-1}\mathbf{C}_h \mathbf{C}_e \mathbf{x}_2 = \xi \Delta t \varepsilon^{-1}\mathbf{C}_h \mathbf{x}_1. \tag{3.48}$$

This is then substituted into the first term of the second row, leading to:

$$\left(\xi^2 + \xi c^2 \Delta t^2 \mathbf{M} + c^2 \Delta t^2 \mathbf{M} \right) \mathbf{x}_2 = 0 \tag{3.49}$$

where

$$\mathbf{M} = \mathbf{C}_h \mathbf{C}_e \tag{3.50}$$

which represents the discrete curl-curl operator. It can be shown that the sparse matrix $\mathbf{M}$ is positive-definite and symmetric. As a result, it is diagonalizable. A projection matrix $\mathbf{P}$ which diagonalizes $\mathbf{M}$ can be introduced, such that:

$$\mathbf{P}^{-1}\mathbf{M}\mathbf{P} = \mathbf{D}_m \tag{3.51}$$

where $\mathbf{D}_m$ is a diagonal matrix containing the eigenvalues of $\mathbf{M}$. This transformation is applied to (3.49):

$$\mathbf{P}^{-1}\left(\xi^2 + \xi c^2 \Delta t^2 \mathbf{M} + c^2 \Delta t^2 \mathbf{M} \right) \mathbf{P}\mathbf{x}_2 = 0 \tag{3.52}$$

$$\left(\mathbf{I}\xi^2 + \xi c^2 \Delta t^2 \mathbf{D}_m + c^2 \Delta t^2 \mathbf{D}_m \right) \mathbf{x}_2 = 0. \tag{3.53}$$

A necessary condition for (3.53) to be zero is then:

$$\left(\xi^2 + \xi c^2 \Delta t^2 \lambda_m + c^2 \Delta t^2 \lambda_m \right) = 0 \tag{3.54}$$

where λ_m are the eigenvalues of $\mathbf{M}$. Solving for ξ from (3.54), leads to:

$$\xi = -\frac{c^2 \Delta t^2}{2}\lambda_m \pm \sqrt{\left(\frac{c^2 \Delta t^2 \lambda_m}{2} \right)^2 - 2\left(\frac{c^2 \Delta t^2 \lambda_m}{2} \right)}. \tag{3.55}$$

Since $\xi = \lambda_G - 1$, the eigenvalues of $\mathbf{G}$ are determined to be:

$$\lambda_G = 1 - \frac{c^2 \Delta t^2}{2} \lambda_m \pm \sqrt{\left(\frac{c^2 \Delta t^2 \lambda_m}{2}\right)^2 - 2\left(\frac{c^2 \Delta t^2 \lambda_m}{2}\right)}. \qquad (3.56)$$

As stated earlier, stability requires $|\lambda_G| \le 1$. From (3.56) this will be true if:

$$\begin{aligned} &i)\ \lambda_m \text{ are real and distinct, and} \\ &ii)\ c^2 \Delta t^2 \lambda_m / 2 \le 2 \end{aligned} \qquad (3.57)$$

Since $\mathbf{M}$ is a positive definite matrix, then λ_m are positive real and distinct. The second condition then constrains the time-step based on the *maximum* eigenvalue of $\mathbf{M}$. It can be shown that for a uniformly spaced rectangular grid, the eigenvalues of $\mathbf{M}$ have the property (c.f., Problem 3, Section 3.10):

$$\lambda_m \le 4\left(\frac{1}{\Delta x^2} + \frac{1}{\Delta y^2} + \frac{1}{\Delta z^2}\right). \qquad (3.58)$$

Applying the eigenvalues λ_m to condition *ii)* of (3.57), it is found that the time-step is bound by the limit:

$$\Delta t \le \frac{1}{c} \cdot \frac{1}{\sqrt{\frac{1}{\Delta x^2} + \frac{1}{\Delta y^2} + \frac{1}{\Delta z^2}}}. \qquad (3.59)$$

As expressed, this limit enforces stability weakly. That is, allowing the possibility of an equality in (3.59) pushes the algorithm to the edge of the stability limit. It is possible in some instances that numerical rounding errors could lead to late time stability. Rather, a strict stability limit is enforced as:

$$\Delta t < \frac{1}{c} \cdot \frac{1}{\sqrt{\frac{1}{\Delta x^2} + \frac{1}{\Delta y^2} + \frac{1}{\Delta z^2}}}. \qquad (3.60)$$

This defines the Courant-Fredrichs-Lewy (CFL) stability limit of the three-dimensional Yee-algorithm for Maxwell's equations [4]. This limit restricts the choice of the time-step. As with the one-dimensional problem, it is found that the limit restricts the distance of propagation per time-step across a unit cell. As an example, consider a cubic grid, where $\Delta x = \Delta y = \Delta z = \Delta$, the stability limit becomes:

$$\Delta t < \frac{\Delta}{c} \frac{1}{\sqrt{3}}. \qquad (3.61)$$

Thus, a wave propagating through the three-dimensional mesh is restricted to propagate only a fraction of the distance across a cell, which is one-third of the distance across the cubic cell diagonal.

One question to ponder is what happens if the media is inhomogeneous, and c is no longer a constant. In this case, it is the largest wave speed that restricts the time-step. The largest wave speed occurs in the medium with the smallest material constants. Thus, in general, one can express:

$$\Delta t < \frac{1}{c_{\max}} \frac{1}{\sqrt{\frac{1}{\Delta x^2} + \frac{1}{\Delta y^2} + \frac{1}{\Delta z^2}}} \qquad (3.62)$$

where c_{max} is the maximum wave speed in the problem domain.

Unlike the one-dimensional case, there is no cancellation of error leading to a "magic time-step." Consequently, the three-dimensional solution will always have discretization error. The properties of this error is defined in the next section. However, it is found that the error is minimized by setting the time-step as close to the CFL-limit as possible. Therefore, one typically chooses:

$$\Delta t = \frac{1}{c_{max}} \frac{\text{CFLN}}{\sqrt{\frac{1}{\Delta x^2} + \frac{1}{\Delta y^2} + \frac{1}{\Delta z^2}}} \quad (3.63)$$

where CFLN is referred to as the "CFL-number." For stability, CFLN < 1. In practice, the CFLN is often chosen to be very close to 1 (e.g., CFLN = 0.99).

3.6 NUMERICAL DISPERSION AND GROUP DELAY

The explicit update equations derived via Yee's algorithm in (3.25)–(3.30) provide an approximate solution to the electromagnetic fields that is expected to converge with second-order accuracy, providing that the CFL-limit of (3.62) is satisfied. The behavior of the error can be better understood by studying the error encountered by a plane wave that is propagating through the discrete space. Assume a uniform monochromatic plane wave propagating through the discrete space, defined with electric and magnetic field vectors:

$$\vec{E} = \vec{E}_o e^{-j\left(\tilde{k}_x x + \tilde{k}_y y + \tilde{k}_z z\right)} e^{j\omega t}, \quad \vec{H} = \vec{H}_o e^{-j\left(\tilde{k}_x x + \tilde{k}_y y + \tilde{k}_z z\right)} e^{j\omega t} \quad (3.64)$$

where $\tilde{k}$ represents the numerical wave number of the discrete space, and ω is the radial frequency of the plane wave. The plane wave must satisfy the discrete form of Maxwell's equations expressed in (3.21)–(3.24). For example, (3.64) can be inserted into (3.21), leading to:

$$\mu H_{x_o} \left(\frac{e^{j\omega\frac{1}{2}\Delta t} - e^{-j\omega\frac{1}{2}\Delta t}}{\Delta t} \right) e^{-j(\tilde{k}_x i\Delta x + \tilde{k}_y(j+\frac{1}{2})\Delta y + \tilde{k}_z(k+\frac{1}{2})\Delta z)} e^{j\omega(n+\frac{1}{2})\Delta t} =$$

$$\left[E_{y_o} \left(\frac{e^{-j\tilde{k}_z\frac{1}{2}\Delta z} - e^{+j\tilde{k}_z\frac{1}{2}\Delta z}}{\Delta z} \right) - E_{z_o} \left(\frac{e^{-j\tilde{k}_y\frac{1}{2}\Delta y} - e^{j\tilde{k}_y\frac{1}{2}\Delta y}}{\Delta y} \right) \right]$$

$$e^{-j(\tilde{k}_x i\Delta x + \tilde{k}_y(j+\frac{1}{2})\Delta y + \tilde{k}_z(k+\frac{1}{2})\Delta z)} e^{j\omega(n+\frac{1}{2})\Delta t}. \quad (3.65)$$

The exponentials on either side of the equation cancel. Applying Euler's formula, this can be expressed as:

$$\frac{\mu}{\Delta t} H_{x_o} \sin\left(\frac{\omega\Delta t}{2}\right) = \left[-\frac{E_{y_o}}{\Delta z} \sin\left(\frac{\tilde{k}_z\Delta z}{2}\right) + \frac{E_{z_o}}{\Delta y} \sin\left(\frac{\tilde{k}_y\Delta y}{2}\right) \right]. \quad (3.66)$$

Let

$$S_t = \frac{1}{\Delta t} \sin\left(\frac{\omega\Delta t}{2}\right), \quad \text{and } S_\kappa = \frac{1}{\Delta\kappa} \sin\left(\frac{\tilde{k}_\kappa\Delta\kappa}{2}\right), \quad (\kappa = x, y, z). \quad (3.67)$$

Then, (3.66) can be written with reduced notation as:

$$H_{x_o} = \frac{1}{\mu S_t} \left[-E_{y_o} S_z + E_{z_o} S_y \right].$$

(3.68)

Similarly,

$$H_{y_o} = \frac{1}{\mu S_t} \left[-E_{z_o} S_x + E_{x_o} S_z \right]$$

(3.69)

$$H_{z_o} = \frac{1}{\mu S_t} \left[-E_{x_o} S_y + E_{y_o} S_x \right].$$

(3.70)

Combining (3.68)–(3.70), these equations can be expressed in vector notation as:

$$\vec{H} = \frac{\vec{S} \times \vec{E}}{\mu S_t}$$

(3.71)

where $\vec{S} = \hat{x} S_x + \hat{y} S_y + \hat{z} S_z$. Dually, from Ampère's law:

$$\vec{E} = -\frac{\vec{S} \times \vec{H}}{\varepsilon S_t}.$$

(3.72)

Inserting (3.72) into (3.71) leads to the discrete vector wave equation:

$$\mu \varepsilon S_t^2 \vec{H} + \vec{S} \times \vec{S} \times \vec{H} = 0.$$

(3.73)

A dual expression can be derived for the electric field vector. As discussed in Section 3.3, the discrete fields also satisfy Gauss's law. For the discrete plane wave, the discrete form of Gauss's law in (3.33) can be expressed as:

$$\mu S_t \left(S_x H_{x_o} + S_y H_{y_o} + S_z H_{z_o} \right) = 0.$$

(3.74)

Consider the x-projection of (3.73):

$$\frac{1}{c^2} S_t^2 H_{x_o} - \left(S_y^2 + S_z^2 \right) H_{x_o} + S_x S_y H_{y_o} + S_x S_z H_{z_o} = 0.$$

(3.75)

From (3.74), $S_y H_{y_o} + S_z H_{z_o} = -S_x H_{x_o}$. Therefore, (3.75) becomes:

$$\frac{1}{c^2} S_t^2 H_{x_o} - \left(S_y^2 + S_z^2 + S_x^2 \right) H_{x_o} = 0.$$

(3.76)

This equality is satisfied iff $H_{x_o} = 0$, or

$$\frac{1}{c^2} S_t^2 = \left(S_x^2 + S_y^2 + S_z^2 \right).$$

(3.77)

This exercise is repeated for all projections of $\vec{H}$ and $\vec{E}$. It is found that (3.77) must be satisfied by all six projections. In conclusion, (3.77) must be satisfied for any non-zero plane wave propagating through the medium. This equation is referred to as the *dispersion relation* that governs a plane wave propagating through the discrete space.

The dispersion relation is important since it governs the propagation behavior of the wave. Since the S's are a function of the space and time-discretization, it is observed that these also impact the wave propagation in a manner that is non-physical. Consequently, studying the dispersion relation will help to understand the properties of the errors resulting from the discrete representation of Maxwell's equations.

Initially, consider the condition where $\lim \Delta x, \Delta y, \Delta z \to 0$ and, consequently, $\lim \Delta t \to 0$ from (3.63). In this circumstance, (3.77) reduces to:

$$\frac{\omega^2}{c^2} = \left(\tilde{k}_x^2 + \tilde{k}_y^2 + \tilde{k}_z^2 \right) \tag{3.78}$$

which is the exact dispersion relation governing the wave-equation in a continuous space [1]. Thus, we can conclude that in the limit that the discrete cell samples tends to 0, the discrete approximation converges to the exact answer. This is a comforting fact!

Next, consider the case of a general discretization. The objective here is compute the error in the numerical wave number, relative to the exact wave number, as a function of the space discretization, the frequency, and the propagation direction through the grid. For the latter, the propagation angle is specified as a spherical angle pair (θ, ϕ), where θ is the zenith angle, and ϕ is the azimuthal angle, such that:

$$\tilde{k}_x = \tilde{k} \cos \phi \sin \theta, \ \ \tilde{k}_y = \tilde{k} \sin \phi \sin \theta, \ \ \tilde{k}_z = \tilde{k} \cos \theta \tag{3.79}$$

where $\tilde{k}$ is the numerical wave number. The objective is to solve for $\tilde{k}$ as a function of $\Delta x, \Delta y, \Delta z, \Delta t$, (θ, ϕ), and ω.

The numerical solution for $\tilde{k}$ is computed via (3.77). This requires minimizing the functional:

$$f\left(\tilde{k}\right) = \frac{1}{c^2} S_t^2 - \left(S_x^2 + S_y^2 + S_z^2 \right) \tag{3.80}$$

via a non-linear solution method. This is done here using Newton's method [9], which solves for $\tilde{k}$ using the iterative solution:

$$\tilde{k}_{i+1} = \tilde{k}_i - \frac{f(\tilde{k}_i)}{f'(\tilde{k}_i)} \tag{3.81}$$

where the index i is the iteration number and f' is the derivative of f with respect to $\tilde{k}$. The initial guess ($i = 0$) is chosen to be the exact solution: $\tilde{k}_0 = \omega/c$. Convergence is obtained in the limit that $\left| f(\tilde{k}_i) \big/ f'(\tilde{k}_i) \right| < tol$, where *tol* is a desired error tolerance. Following this procedure, (3.81) typically converges quite rapidly.

For practical sampling, the numerical wave number is purely real. As a consequence, the error is purely dispersive. That is, the error results in an effective phase delay of the wave that is frequency dependent. For very coarse discretizations (near the Nyquist limit, or larger), the wave number is complex. Thus, the waves encounter a large attenuation and the discrete medium is highly dissipative. This is actually to our advantage in that high frequency errors will tend to be damped out. At this point, the focus is on the propagation characteristics of signals within the desired frequency band.

The exact wave number is:

$$k = \frac{\omega}{c} = \frac{2\pi f}{c} = \frac{2\pi}{\lambda} \tag{3.82}$$

where λ is the wavelength (a wavelength is the distance traveled by the wave in which it undergoes a full 2π radian (or 360-degree) phase shift). When $\tilde{k}$ differs from the exact wave number, it translates to wave number a phase error accumulated by the wave propagating over distance through the discrete space. The phase error can be measured in degrees per wavelength as:

$$\text{phase error}(°/\lambda) = \frac{180°}{\pi} \cdot \left(\tilde{k} - k \right). \tag{3.83}$$

Figure 3.3 illustrates a polar plot of the phase error as a function of the spherical angles (θ, ϕ) when $\Delta x = \Delta y = \Delta z = 0.05\lambda$ (or 20 divisions per wavelength), and for a CFLN = 0.99. The maximum phase error is roughly 1° per wavelength for this sampling. It is seen that the phase error

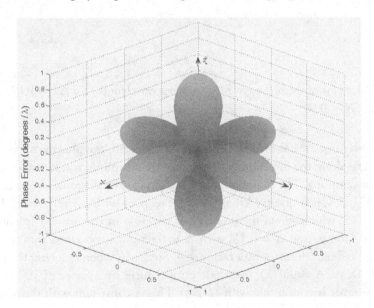

Figure 3.3: Polar plot of the phase error (degrees per wavelength) for $\Delta x = \Delta y = \Delta z = 0.05\lambda$, and CFLN = 0.99.

is strongly a function of angle. It is observed that the maximum phase error occurs when the wave is propagating along the cardinal axes. The minimum error occurs when the wave is traveling close to the diagonal (e.g., close to $(\theta, \phi) = (45°, 45°)$). This variation in the phase error implies that the effective medium of the discrete space has an anisotropic behavior. Figure 3.4 illustrates a polar plot of the phase error for a non cubic grid. In this case, $\Delta x = 0.05\lambda$, $\Delta y = 0.04\lambda$, and $\Delta z = 0.03\lambda$. This

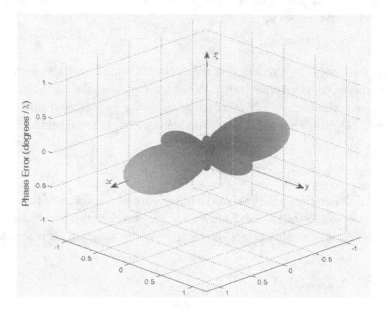

Figure 3.4: Polar plot of the phase error (degrees per wavelength) for $\Delta x = 0.05\lambda$, $\Delta y = 0.04\lambda$, $\Delta z = 0.03\lambda$, and CFLN = 0.99.

leads to a non-symmetric anisotropic behavior. As expected, the phase error is maximum when the wave is traveling along the direction with the coarsest sampling, and is reduced along the direction with the finest sampling. The minimum error is still near the diagonal of the mesh.

From Fig. 3.4, it is observed that the error reduces with discretization. This is further illustrated by Fig. 3.5, which shows the phase error versus ϕ in the $\theta = 90°$ plane for different cubic discretizations. Comparing $\Delta = 0.1\lambda$, with $\Delta = 0.05\lambda$, it is observed that reducing the sampling by a factor of 2 reduces the error by a factor of 4. Similarly, when reducing the sampling by a factor of 10 ($\Delta = 0.1\lambda$ vs. $\Delta = 0.01\lambda$), the phase error reduces by a factor of 100. Thus, it is concluded that the phase error is converging with $O(\Delta^2)$. This is consistent with the second-order accuracy expected from the central difference approximation. Also, what is also gleaned from this plot is that a phase error of $< 1°/\lambda$ requires a special sampling of $\lambda/20$. It should be realized that as the wave propagates over a distance, phase error is accumulated. Consequently, for very large problems, finer discretizations may be necessary in order to reduce the global solution error.

Figure 3.6 illustrates the phase error as a function of ϕ in the $\theta = 90°$ plane for different values of the CFLN. What is observed is that by decreasing the CFLN, the error actually increases! While this seems somewhat counter-intuitive, the reason for this is that there is some added benefit

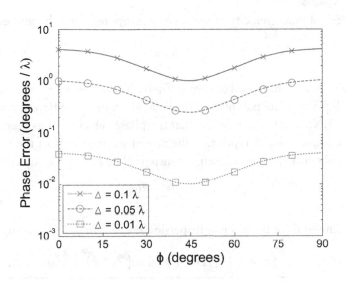

Figure 3.5: Phase error (degrees per wavelength) for $\Delta x = \Delta y = \Delta z = \Delta$ with CFLN $= 0.99$.

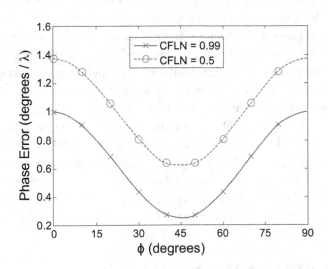

Figure 3.6: Phase error (degrees per wavelength) for $\Delta x = \Delta y = \Delta z = 0.005\lambda$ for different CFLN in the $\theta = 90°$ plane.

of cancellation of the time and space discretization errors, which is maximized when the CFLN is as close to 1 as possible.

The numerical phase velocity of the wave is computed from the numerical wave number:

$$\tilde{v}_p = \frac{\omega}{\tilde{k}}. \tag{3.84}$$

The exact phase velocity is the speed of light c. Figure 3.7 (a) illustrates the phase velocity normalized by the speed of light as a function of ϕ in the $\theta = 90°$ plane for different cubic discretizations and for a reduced CFLN. What is observed is that the phase velocity is less than the speed of light.

While the phase velocity represents the rate of advancement of the phase over distance, the group velocity is actually the rate of energy transport. The group velocity is measured as:

$$v_g = \frac{d\omega}{dk}. \tag{3.85}$$

This can be computed from the discrete dispersion relation in (3.77), leading to:

$$\tilde{v}_g = \frac{c^2}{\omega \tilde{k}} \frac{\left(\tilde{k}_x^2 \mathrm{sinc}\left(\tilde{k}_x \Delta x \right) + \tilde{k}_y^2 \mathrm{sinc}\left(\tilde{k}_y \Delta y \right) + \tilde{k}_z^2 \mathrm{sinc}\left(\tilde{k}_z \Delta z \right) \right)}{\mathrm{sinc}\left(\omega \Delta t \right)} \tag{3.86}$$

where $\mathrm{sinc}(x) = \sin(x)/x$. Figure 3.7 (b) illustrates the numerical group velocity normalized by the speed of light for the same discretizations illustrated in Fig. 3.7 (a). Similar to the phase velocity, the group velocity is also a function of discretization. It also converges with second-order accuracy. The error in the group velocity is also minimized by maximizing the time-step. Comparing Fig. 3.7 (b) with Fig. 3.7 (a), it is observed that the group velocity suffers from slightly worse error than the phase velocity. However, it is of the same order of magnitude. Finally, this study involved a lossless, linear, isotropic material. For this material, one would expect $v_g \cdot v_p = c^2$. In the discrete FDTD space, $v_g \cdot v_p < c^2$, indicating that the error is non-physical in nature.

In summary, the discrete FDTD space is inherently dispersive. That is, the phase and group velocities are frequency dependent. It is also anisotropic since the phase and group velocities are a function of the angle which the wave is propagating through the medium. In order to minimize the error to less than $1°/\lambda$, a space discretization of $\Delta/\lambda \leq 20$ is recommended, where λ represents the wavelength of the maximum frequency of the signal bandwidth. Finally, it is also found that the time-step should be chosen as close to the CFL limit as possible. Typically, CFLN = 0.99 is an excellent choice.

3.7 MATERIAL AND BOUNDARIES

The discrete fields of Yee's FDTD algorithm are subject to the boundary conditions posed in (3.11)–(3.18). How these boundary conditions are satisfied in the discrete FDTD space is the topic of this section. To begin, consider a surface separating two dielectric materials. The geometry is discretized such that surface is defined by primary grid faces, as illustrated in Fig. 3.8. The discrete electric field

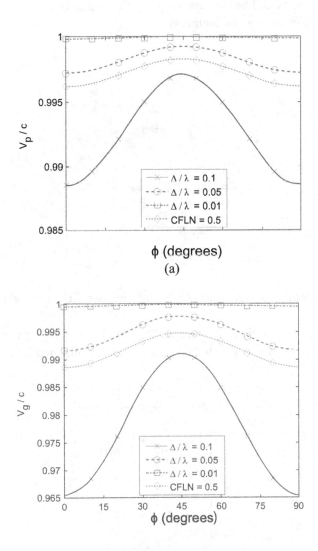

Figure 3.7: (a) Numerical phase velocity, and (b) numerical group velocity as a function of ϕ in the $\theta = 90°$ plane. Both are computed for $\Delta x = \Delta y = \Delta z = \Delta$ with CFLN = 0.99, and for CFLN = 0.5 with $\Delta = 0.05\lambda$.

vectors are sampled in the plane of S and are purely tangential to S. Since the electric field edges are shared by the cells above and below the faces, the discretization naturally enforces the continuity of the tangential electric field. The discrete magnetic flux is purely normal to S, and is also inherently forced to be continuous across the boundary.

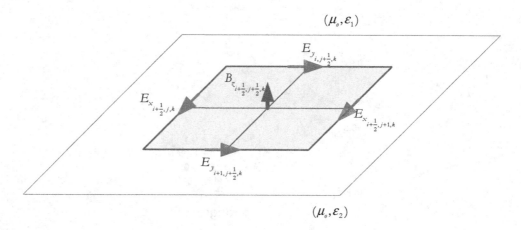

Figure 3.8: Primary grid cell face on a surface separating two dielectric regions.

Next, consider the updates of the electric field vectors tangent to the plane S. To illustrate this, consider the secondary grid face in Fig. 3.9, which is bisected by S. Note that the dielectric

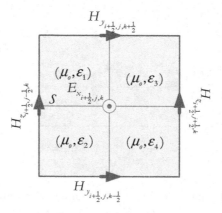

Figure 3.9: Secondary grid face with center edge shared by multiple media.

media interface is split along the horizontal plane. For further generality, it is assumed that just outside of the primary cell face of Fig. 3.8 are two other media, ε_3 and ε_4, separated by the same

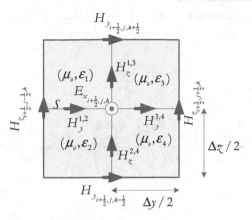

Figure 3.10: Secondary grid face split into four quadrants.

horizontal plane. The four dielectric regions share the primary grid cell edge of $E_{x_{i+\frac{1}{2},j,k}}$. The update of the electric field is computed via (3.28). The quandary that arises is how to evaluate ε to properly represent the constitutive relationship since the media is inhomogeneous. To do this, the face is split into four quadrants, as illustrated in Fig. 3.10. Based on the fundamental Yee-discretization, E_x is assumed to be homogeneous over all four quadrants. This maintains the continuity of E_x across all four interfaces. The tangential H-fields on the perimeter edges are assumed to be constant along their respective edges as well. Since μ is held constant, this is a valid assumption. Next, interior edges are introduced on the boundaries separating the dielectric regions. Tangential H-fields are placed on these interior edges. Finally, Ampère's law in its *integral* form is enforced over each of the four quadrants. For example, in quadrant 1:

$$\varepsilon_1 \frac{E_x^{n+\frac{1}{2}}_{i+\frac{1}{2},j,k} - E_x^{n-\frac{1}{2}}_{i+\frac{1}{2},j,k}}{\Delta t} \frac{\Delta z \Delta y}{4} = H_y^{1,2^n} \frac{\Delta y}{2} + H_z^{1,3^n} \frac{\Delta z}{2} - H_{y_{i+\frac{1}{2},j,k+\frac{1}{2}}}^{n} \frac{\Delta y}{2} - H_{z_{i+\frac{1}{2},j-\frac{1}{2},k}}^{n} \frac{\Delta z}{2}.$$

(3.87)

This is repeated in the other three quadrants. Finally, the discrete Ampère's law of all four quadrants are summed, leading to:

$$\frac{\varepsilon_1 + \varepsilon_2 + \varepsilon_3 + \varepsilon_4}{4} \frac{E_x^{n+\frac{1}{2}}_{i+\frac{1}{2},j,k} - E_x^{n-\frac{1}{2}}_{i+\frac{1}{2},j,k}}{\Delta t} \Delta z \Delta y =$$

$$H_{y_{i+\frac{1}{2},j,k-\frac{1}{2}}}^{n} \Delta y + H_{z_{i+\frac{1}{2},j+\frac{1}{2},k}}^{n} \Delta z - H_{y_{i+\frac{1}{2},j,k+\frac{1}{2}}}^{n} \Delta y - H_{z_{i+\frac{1}{2},j-\frac{1}{2},k}}^{n} \Delta z$$

(3.88)

where the interior magnetic fields cancel due to complementary signs. Dividing both sides by $\Delta z \Delta y$, this leads to the recursive update expression:

$$
E_x{}^{n+\frac{1}{2}}_{i+\frac{1}{2},j,k} = E_x{}^{n-\frac{1}{2}}_{i+\frac{1}{2},j,k}
$$

$$
+ \frac{\Delta t}{\varepsilon_{i+\frac{1}{2},j,k}} \left[\left(\frac{H_z{}^n_{i+\frac{1}{2},j+\frac{1}{2},k} - H_z{}^n_{i+\frac{1}{2},j-\frac{1}{2},k}}{\Delta y} \right) - \left(\frac{H_y{}^n_{i+\frac{1}{2},j,k+\frac{1}{2}} - H_y{}^n_{i+\frac{1}{2},j,k-\frac{1}{2}}}{\Delta z} \right) \right] \tag{3.89}
$$

where

$$
\varepsilon_{i+\frac{1}{2},j,k} = \frac{\varepsilon_1 + \varepsilon_2 + \varepsilon_3 + \varepsilon_4}{4} \tag{3.90}
$$

is the average permittivity of the secondary grid face.

Next, consider the dual problem, that is, a surface separating two magnetic material regions. To properly enforce the continuity of the tangential magnetic field across the boundary, the surface must be defined by faces of the *secondary* grid. The normal electric flux density will also be inherently continuous. Since the secondary grid faces define magnetic material boundaries, the magnetic material will be constant over a secondary grid cell face, which was the pre-requisite for the proof of (3.89). Similarly, the dielectric material must be held constant over a primary grid face. Consequently, if a material has both electric and magnetic properties, then the material boundary will have to be staggered by $1/2$ a grid cell in order to maintain the appropriate field continuities. This staggering leads to discretization error that is unavoidable with the staggered grid approximation.

In summary, when modeling inhomogeneous media, the material constants will be a function of position and will be averaged as in (3.90). That is the local permittivity of an edge is the average permittivity of the four primary grid cells sharing the edge. Similarly, the local permeability of a secondary grid edge is the average permeability of the four secondary grid cells sharing the edge. This formulation is stable within the CFL stability limit (3.62), where $c_{\max}$ is the maximum propagation speed, which corresponds to the region with the minimum μ and ε.

Next, consider a perfect electrical conductor (PEC) boundary. On the surface of the PEC, the fields satisfy the boundary conditions specified in (3.15) and (3.16). Since the tangential electric fields are zero on the surface of a PEC boundary, it is natural to define the boundary with primary grid cell faces. For example, consider Fig. 3.8 with region 2 being a PEC volume. The tangential fields on the perimeter of the primary grid face lie on S_{PEC}. Consequently, these fields are forced to be zero over all time. This boundary condition is referred to as a *Dirichlet* boundary condition. It is also observed from (3.27), that the magnetic field normal to S_{PEC} will implicitly be zero since the electric fields are zero (H_z must also have an initial value of 0). In the staggered space, neither the normal electric field or the tangential magnetic fields are sampled on S_{PEC}, and thus are updated as normal.

A perfect magnetic conductor (PMC) boundary can be modeled in a dual manner by modeling the PMC boundary by secondary grid faces. However, PMC materials are non-physical, and typically PMC planes are used to enforce symmetry about a plane to reduce the overall problem dimension. Since the global problem space is typically represented via the primary grid, it becomes necessary to be able to model a PMC plane via primary grid faces. Consider again Fig. 3.8 with region 2 now being a PMC volume, and the surface S now being a PMC surface: S_{PMC}. In this situation, the boundary conditions (3.17) and (3.18) should be satisfied. On a PMC plane, it is observed that the tangential H-field or the normal D-field is zero. However, this cannot be strictly enforced by the staggered grid since neither of these projections are sampled on the plane. Rather, the tangential electric fields are specified on the plane. The difficulty arises when attempting to update the electric fields on the PMC plane since the normal derivatives of the tangent H-fields cannot be computed. This can be handled via image theory. For example, consider the updates of E_x and E_y on the z-normal PMC plane via (3.28) and (3.29). Applying image theory, the discrete tangential H-fields on either side of the PMC plane $z = k\Delta z$ satisfy the relationships:

$$H^n_{y_{i+\frac{1}{2},j,k-\frac{1}{2}}} = -H^n_{y_{i+\frac{1}{2},j,k+\frac{1}{2}}}, \quad H^n_{x_{i,j+\frac{1}{2},k-\frac{1}{2}}} = -H^n_{x_{i,j+\frac{1}{2},k+\frac{1}{2}}}. \tag{3.91}$$

As a consequence, the updates of E_x and E_y on the PMC plane would be:

$$E^{n+\frac{1}{2}}_{x_{i+\frac{1}{2},j,k}} = E^{n-\frac{1}{2}}_{x_{i+\frac{1}{2},j,k}} + \frac{\Delt t}{\varepsilon}\left[\left(\frac{H^n_{z_{i+\frac{1}{2},j+\frac{1}{2},k}} - H^n_{z_{i+\frac{1}{2},j-\frac{1}{2},k}}}{\Delta y}\right) - \left(\frac{2H^n_{y_{i+\frac{1}{2},j,k+\frac{1}{2}}}}{\Delta z}\right)\right], \tag{3.92}$$

$$E^{n+\frac{1}{2}}_{y_{i,j+\frac{1}{2},k}} = E^{n-\frac{1}{2}}_{y_{i,j+\frac{1}{2},k}} + \frac{\Delta t}{\varepsilon}\left[\left(\frac{2H^n_{x_{i,j+\frac{1}{2},k+\frac{1}{2}}}}{\Delta z}\right) - \left(\frac{H^n_{z_{i+\frac{1}{2},j+\frac{1}{2},k}} - H^n_{z_{i-\frac{1}{2},j+\frac{1}{2},k}}}{\Delta x}\right)\right]. \tag{3.93}$$

If the PMC region were volume 1 such that the region above the plane S_{PMC} is PMC, then these update expressions would instead be expressed as:

$$E^{n+\frac{1}{2}}_{x_{i+\frac{1}{2},j,k}} = E^{n-\frac{1}{2}}_{x_{i+\frac{1}{2},j,k}} + \frac{\Delta t}{\varepsilon}\left[\left(\frac{H^n_{z_{i+\frac{1}{2},j+\frac{1}{2},k}} - H^n_{z_{i+\frac{1}{2},j-\frac{1}{2},k}}}{\Delta y}\right) + \left(\frac{2H^n_{y_{i+\frac{1}{2},j,k-\frac{1}{2}}}}{\Delta z}\right)\right], \tag{3.94}$$

$$E^{n+\frac{1}{2}}_{y_{i,j+\frac{1}{2},k}} = E^{n-\frac{1}{2}}_{y_{i,j+\frac{1}{2},k}} + \frac{\Delta t}{\varepsilon}\left[\left(\frac{-2H^n_{x_{i,j+\frac{1}{2},k-\frac{1}{2}}}}{\Delta z}\right) - \left(\frac{H^n_{z_{i+\frac{1}{2},j+\frac{1}{2},k}} - H^n_{z_{i-\frac{1}{2},j+\frac{1}{2},k}}}{\Delta x}\right)\right]. \tag{3.95}$$

The updates in (3.92)–(3.95) are stable within the stability limit.

3.8   LOSSY MEDIA

In most physical problems, there is a small amount of loss. Often, the loss can be modeled as conductor losses. For example, given a finite-conductivity, Ampère's law is written as:

$$\frac{\partial \vec{D}}{\partial t} + \sigma \vec{E} = \nabla \times \vec{H} \tag{3.96}$$

where $\sigma \vec{E}$ is the loss term. This is discretized using the standard Yee-algorithm. For example, consider the x-projection:

$$\varepsilon \left(\frac{E_x^{n+\frac{1}{2}}\big|_{i+\frac{1}{2},j,k} - E_x^{n-\frac{1}{2}}\big|_{i+\frac{1}{2},j,k}}{\Delta t} \right) + \sigma E_x^n\big|_{i+\frac{1}{2},j,k} = \\ \left(\frac{H_z^n\big|_{i+\frac{1}{2},j+\frac{1}{2},k} - H_z^n\big|_{i+\frac{1}{2},j-\frac{1}{2},k}}{\Delta y} \right) - \left(\frac{H_y^n\big|_{i+\frac{1}{2},j,k+\frac{1}{2}} - H_y^n\big|_{i+\frac{1}{2},j,k-\frac{1}{2}}}{\Delta z} \right) \tag{3.97}$$

The problem is that since the electric field is staggered in time, the loss term σE_x^n cannot be evaluated explicitly at time $t = n\Delta t$. One way to circumvent this is to "time-average" σE_x^n, using known values at times $t = (n + \frac{1}{2}\Delta t)$ and $t = (n - \frac{1}{2}\Delta t)$. A linear averaging is second-order accurate and is commensurate with the accuracy of the central difference operators. This leads to:

$$\varepsilon \left(\frac{E_x^{n+\frac{1}{2}}\big|_{i+\frac{1}{2},j,k} - E_x^{n-\frac{1}{2}}\big|_{i+\frac{1}{2},j,k}}{\Delta t} \right) + \sigma \left(\frac{E_x^{n+\frac{1}{2}}\big|_{i+\frac{1}{2},j,k} + E_x^{n-\frac{1}{2}}\big|_{i+\frac{1}{2},j,k}}{2} \right) = \\ \left(\frac{H_z^n\big|_{i+\frac{1}{2},j+\frac{1}{2},k} - H_z^n\big|_{i+\frac{1}{2},j-\frac{1}{2},k}}{\Delta y} \right) - \left(\frac{H_y^n\big|_{i+\frac{1}{2},j,k+\frac{1}{2}} - H_y^n\big|_{i+\frac{1}{2},j,k-\frac{1}{2}}}{\Delta z} \right) \tag{3.98}$$

Grouping terms, the explicit update expression is derived as:

$$E_x^{n+\frac{1}{2}}\big|_{i+\frac{1}{2},j,k} = \frac{\frac{\varepsilon}{\Delta t} - \frac{\sigma}{2}}{\frac{\varepsilon}{\Delta t} + \frac{\sigma}{2}} E_x^{n-\frac{1}{2}}\big|_{i+\frac{1}{2},j,k} + \\ \frac{1}{\frac{\varepsilon}{\Delta t} + \frac{\sigma}{2}} \left(\frac{H_z^n\big|_{i+\frac{1}{2},j+\frac{1}{2},k} - H_z^n\big|_{i+\frac{1}{2},j-\frac{1}{2},k}}{\Delta y} \right) - \left(\frac{H_y^n\big|_{i+\frac{1}{2},j,k+\frac{1}{2}} - H_y^n\big|_{i+\frac{1}{2},j,k-\frac{1}{2}}}{\Delta z} \right). \tag{3.99}$$

Similar expressions can be derived for the updates of E_y and E_z. It can be proven using an eigenvalue analysis similar to that in Section 3.5 that this updating scheme is stable within the CFL stability limit (3.62). An important condition for this is that the coefficient weighting $E_x^{n-\frac{1}{2}}$ must be ≤ 1. Thus, stability requires that σ be a non-negative real value.

If the lossy medium is inhomogeneous, then one can follow the proof in Section 3.7 to show that regions of different conductivity are separated by surfaces represented by the primary grid faces. The conductivity is then averaged in an identical manner as the permittivity following (3.90). This scheme preserves the proper field continuities, and it maintains the second-order accuracy of the FDTD algorithm.

3.9 DISPERSIVE MEDIA

Dispersive materials are materials with relative permeabilities or permittivities that are frequency dependent. Such materials have been modeled with FDTD and FETD (finite-element time-domain) methods [10]–[12]. Modeling dispersive materials is important to a number of applications. For example, applications in microwave therapy, imaging, or bio-electromagnetic hazards require the simulation of waves in lossy, dispersive biological tissues. Similarly, optical or microwave lenses used for wavelength splitting are also inherently dispersive. Recently, there has been much interest in meta-materials, which, by nature have material constants that are strongly dispersive.

The difficulty of simulating dispersive media in the time domain is that the constitutive relations require a convolution. A direct convolution is usually impractical since it requires storing the full time-history of the fields. More efficient methods have been proposed for FDTD methods, including the discrete recursive convolution method [13, 14] or the auxiliary differential equation method (ADE) [10]–[12], [15]. In this section, the more conventional approach based on the ADE method is presented.

For sake of illustration, consider a linear, isotropic Debye media model that has an electric susceptibility:

$$\chi(\omega) = \sum_{p=1}^{N_p} \frac{\Delta \varepsilon_p}{1 + j\omega\tau_p} \qquad (3.100)$$

where N_p = the number of Debye poles, $\Delta \varepsilon_p = \varepsilon_s - \varepsilon_\infty$ is the change in relative permittivity, ε_s is the static or zero-frequency relative permittivity and ε_∞ is the relative permittivity at infinite frequency, and τ_p = the relaxation time for each pole. The relative permittivity of the material is expressed as:

$$\varepsilon(\omega) = \varepsilon_o (\varepsilon_\infty + \chi(\omega)). \qquad (3.101)$$

The time-rate of change of the electric flux density can be transformed into the frequency domain as:

$$\frac{\partial}{\partial t}\left(\varepsilon(t) * \vec{E}\right) \rightarrow j\omega\varepsilon(\omega)\vec{E} = j\omega\varepsilon_o \left(\varepsilon_\infty + \sum_{p=1}^{N_p} \frac{\Delta \varepsilon_p}{1 + j\omega\tau_p}\right) \vec{E} \qquad (3.102)$$

where $*$ is the convolution operator. An auxiliary electric-polarization vector is introduced for each Debye pole:

$$\vec{P}_p^e(\omega) = \frac{j\omega\Delta\varepsilon_p}{\left(1 + j\omega\tau_p\right)}\vec{E}.$$ (3.103)

Thus, (3.102) can be expressed as:

$$j\omega\varepsilon_o\left(\varepsilon_\infty\vec{E} + \frac{1}{j\omega}\sum_{p=1}^{N_p}\vec{P}_p^e\right) = j\omega\varepsilon_o\varepsilon_\infty\vec{E} + \varepsilon_o\sum_{p=1}^{N_p}\vec{P}_p^e.$$ (3.104)

This is transformed back to the time-domain and then applied within Ampere's law to obtain:

$$\varepsilon_o\varepsilon_\infty\frac{\partial}{\partial t}\vec{E} + \varepsilon_o\sum_{p=1}^{N_p}\vec{P}_p^e = \nabla \times \vec{H}.$$ (3.105)

Equation (3.103), is also transformed to the time-domain, leading to the differential equation:

$$\frac{d}{dt}\vec{P}_p^e + \frac{1}{\tau_p^e}\vec{P}_p^e = \frac{\Delta\varepsilon_p}{\tau_p^e}\frac{d}{dt}\vec{E}.$$ (3.106)

Thus, each $\vec{P}_p^e$ satisfies (3.106), which is referred to as the auxiliary differential equation (ADE).

The set of equations (3.105) and (3.106) is discretized via the FDTD method by applying the standard Yee-algorithm. Observing (3.106), it becomes obvious that the proper discretization for $\vec{P}_p^e$ is to sample it at the edge centers in an identical manner as the electric field. Consider the x-projection of (3.105). This can be cast in a discrete form based on the Yee-algorithm as:

$$\varepsilon_o\varepsilon_\infty\left(\frac{E_x^{n+\frac{1}{2}}\Big|_{i+\frac{1}{2},j,k} - E_x^{n-\frac{1}{2}}\Big|_{i+\frac{1}{2},j,k}}{\Delta t}\right) + \varepsilon_o\sum_{p=1}^{N_p}P_{px}^{e^n}\Big|_{i+\frac{1}{2},j,k} =$$

$$\left(\frac{H_z^n\Big|_{i+\frac{1}{2},j+\frac{1}{2},k} - H_z^n\Big|_{i+\frac{1}{2},j-\frac{1}{2},k}}{\Delta y}\right) - \left(\frac{H_y^n\Big|_{i+\frac{1}{2},j,k+\frac{1}{2}} - H_y^n\Big|_{i+\frac{1}{2},j,k-\frac{1}{2}}}{\Delta z}\right)$$ (3.107)

where the polarization vector is expressed at time-step n to maintain second-order accuracy.

Next, consider the discrete form of (3.105), where the time-derivatives are approximated to second-order accuracy using central differencing:

$$\frac{P_{px}^{e^{n+\frac{1}{2}}}\Big|_{i+\frac{1}{2},j,k} - P_{px}^{e^{n-\frac{1}{2}}}\Big|_{i+\frac{1}{2},j,k}}{\Delta t} + \frac{1}{\tau_p^e}\frac{P_{px}^{e^{n+\frac{1}{2}}}\Big|_{i+\frac{1}{2},j,k} + P_{px}^{e^{n-\frac{1}{2}}}\Big|_{i+\frac{1}{2},j,k}}{2} = \frac{\Delta\varepsilon_p}{\tau_p^e}\frac{E_x^{n+\frac{1}{2}}\Big|_{i+\frac{1}{2},j,k} - E_x^{n-\frac{1}{2}}\Big|_{i+\frac{1}{2},j,k}}{\Delta t}.$$ (3.108)

This can be rewritten in the form of a recursive update equation as:

$$P^{e^{n+\frac{1}{2}}}_{Px_{\,i+\frac{1}{2},j,k}} = \frac{2\tau^e_p - \Delta t}{2\tau^e_p + \Delta t} P^{e^{n-\frac{1}{2}}}_{Px_{\,i+\frac{1}{2},j,k}} + \frac{2\Delta\varepsilon_p}{2\tau^e_p + \Delta t}\left(E^{n+\frac{1}{2}}_{x_{\,i+\frac{1}{2},j,k}} - E^{n-\frac{1}{2}}_{x_{\,i+\frac{1}{2},j,k}}\right). \tag{3.109}$$

Comparing this expression with (3.107), the difficulty arises in that the polarization vector is needed at time-step n. To this end, we approximate:

$$P^{e^{n}}_{Px_{\,i+\frac{1}{2},j,k}} = \frac{1}{2}\left(P^{e^{n+\frac{1}{2}}}_{Px_{\,i+\frac{1}{2},j,k}} + P^{e^{n-\frac{1}{2}}}_{Px_{\,i+\frac{1}{2},j,k}}\right)$$

$$= \frac{2\tau^e_p}{2\tau^e_p + \Delta t} P^{e^{n-\frac{1}{2}}}_{Px_{\,i+\frac{1}{2},j,k}} + \frac{\Delta\varepsilon_p}{2\tau^e_p + \Delta t}\left(E^{n+\frac{1}{2}}_{x_{\,i+\frac{1}{2},j,k}} - E^{n-\frac{1}{2}}_{x_{\,i+\frac{1}{2},j,k}}\right). \tag{3.110}$$

This is then inserted into (3.107), leading to:

$$\varepsilon_o\varepsilon_\infty\left(\frac{E^{n+\frac{1}{2}}_{x_{\,i+\frac{1}{2},j,k}} - E^{n-\frac{1}{2}}_{x_{\,i+\frac{1}{2},j,k}}}{\Delta t}\right) + \varepsilon_o\sum_{p=1}^{N_p}\frac{2\tau^e_p}{2\tau^e_p + \Delta t}P^{e^{n-\frac{1}{2}}}_{Px_{\,i+\frac{1}{2},j,k}} +$$

$$\varepsilon_o\sum_{p=1}^{N_p}\frac{\Delta\varepsilon_p}{2\tau^e_p + \Delta t}\left(E^{n+\frac{1}{2}}_{x_{\,i+\frac{1}{2},j,k}} - E^{n-\frac{1}{2}}_{x_{\,i+\frac{1}{2},j,k}}\right) =$$

$$\left(\frac{H^n_{z_{\,i+\frac{1}{2},j+\frac{1}{2},k}} - H^n_{z_{\,i+\frac{1}{2},j-\frac{1}{2},k}}}{\Delta y}\right) - \left(\frac{H^n_{y_{\,i+\frac{1}{2},j,k+\frac{1}{2}}} - H^n_{y_{\,i+\frac{1}{2},j,k-\frac{1}{2}}}}{\Delta z}\right). \tag{3.111}$$

A recursive update equation can be derived from this, leading to:

$$E^{n+\frac{1}{2}}_{x_{\,i+\frac{1}{2},j,k}} = E^{n-\frac{1}{2}}_{x_{\,i+\frac{1}{2},j,k}} + \frac{1}{\varepsilon_o\left(\frac{\varepsilon_\infty}{\Delta t} + \sum_{p=1}^{N_p}\frac{\Delta\varepsilon_p}{2\tau^e_p + \Delta t}\right)} \cdot$$

$$\left[\left(\frac{H^n_{z_{\,i+\frac{1}{2},j+\frac{1}{2},k}} - H^n_{z_{\,i+\frac{1}{2},j-\frac{1}{2},k}}}{\Delta y}\right) - \left(\frac{H^n_{y_{\,i+\frac{1}{2},j,k+\frac{1}{2}}} - H^n_{y_{\,i+\frac{1}{2},j,k-\frac{1}{2}}}}{\Delta z}\right) - \varepsilon_o\sum_{p=1}^{N_p}\frac{2\tau^e_p}{2\tau^e_p + \Delta t}P^{e^{n-\frac{1}{2}}}_{Px_{\,i+\frac{1}{2},j,k}}\right]$$

$$\tag{3.112}$$

where $P^{e^{n-\frac{1}{2}}}_{Px_{\,i+\frac{1}{2},j,k}}$ satisfies (3.109).

3.10 NON-UNIFORM GRIDDING

The Yee-algorithm assumes a uniform rectangular discretization of the geometry. Unfortunately, the separation of boundaries cannot always be matched by the uniform sampling of the grid. As a consequence, either the boundary location must be shifted, resulting in some error in the solution, or the grid must be made finer until the boundary is matched. Another alternative is to use a non-uniformly spaced grid. This provides some additional flexibility to the Yee-scheme.

A non-uniformly spaced rectangular grid is assumed. Consider the example grid illustrated in Fig. 3.11. The vertices of the mesh are denoted as: (x_i, y_j, z_k). The spacing of the mesh is non-

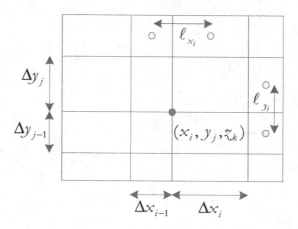

Figure 3.11: Cross-section of the primary non-uniformly spaced grid.

uniform, and thus must be indexed as a function of position. We define the primary grid cell spacing as:

$$\{\Delta x_i = x_{i+1} - x_i\}, \quad \{\Delta y_j = y_{j+1} - y_j\}, \quad \{\Delta z_k = z_{k+1} - z_k\} \tag{3.113}$$

where the indices have the ranges $i = 1, N_x - 1$, $j = 1, N_y - 1$, $k = 1, N_z - 1$, and (N_x, N_y, N_z) is the maximum vertex coordinate. Similar to the uniform grid, a primary grid cell is bound by the primary grid edges. For the non-uniformly spaced grid, cell (i, j, k) will thus have dimensions $(\Delta x_i, \Delta y_j, \Delta z_k)$, which also define the edge and face dimensions.

The secondary grid cells have edges that connect the centers of the primary grid cells. The edges are thus defined by the lengths (c.f., Fig. 3.11):

$$\left\{ \ell_{x_i} = \frac{\Delta x_i + \Delta x_{i-1}}{2} \right\}, \quad \left\{ \ell_{y_j} = \frac{\Delta y_j + \Delta y_{j-1}}{2} \right\}, \quad \left\{ \ell_{z_k} = \frac{\Delta z_k + \Delta z_{k-1}}{2} \right\}. \tag{3.114}$$

Therefore, secondary grid cell (i, j, k) will have dimensions $(\ell_{x_i}, \ell_{y_j}, \ell_{z_k})$.

The discretization of the electric and magnetic fields within the staggered non-uniform grid is defined in a similar manner as the uniformly spaced grid. Tangential projections of the electric fields are sampled at the centers of the primary grid edges and are normal to the secondary grid cell faces. The normal projections of the magnetic field are sampled at the centers of the primary grid cell faces. They are also aligned along the secondary grid cell edges. This is illustrated in Fig. 3.12. The primary grid cell face is illustrated in Fig. 3.12 (a). It is observed that the cell face locally appears

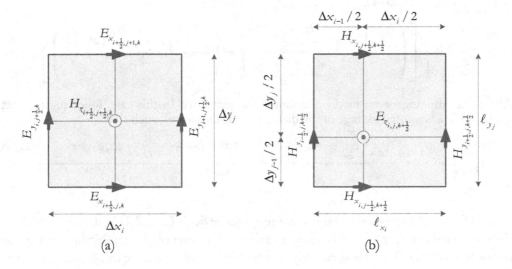

Figure 3.12: Non-uniform grid faces: (a) primary grid face, (b) secondary grid face.

to be identical to a uniformly spaced grid face, with the exception that the face dimensions are a function of discrete position. The update equation for the magnetic field can be derived directly via a central difference approximation, or finite integration, leading to:

$$
H_{z}^{n+1}{}_{i+\frac{1}{2},j+\frac{1}{2},k} = H_{z}^{n}{}_{i+\frac{1}{2},j+\frac{1}{2},k} +
$$

$$
\frac{\Delta t}{\mu_{i+\frac{1}{2},j+\frac{1}{2},k}} \left[\left(\frac{E_{x}^{n+\frac{1}{2}}{}_{i+\frac{1}{2},j+1,k} - E_{x}^{n+\frac{1}{2}}{}_{i+\frac{1}{2},j,k}}{\Delta y_{j}} \right) - \left(\frac{E_{y}^{n+\frac{1}{2}}{}_{i+1,j+\frac{1}{2},k} - E_{y}^{n+\frac{1}{2}}{}_{i,j+\frac{1}{2},k}}{\Delta x_{i}} \right) \right]. \tag{3.115}
$$

For generality, an inhomogeneous permeability is assumed, which again, is the average permeability based on an averaging of the permeability of the four secondary grid cells sharing the secondary grid edge supporting $H_{z}^{n}{}_{i+\frac{1}{2},j+\frac{1}{2},k}$.

Next, consider the secondary grid cell face in Fig. 3.12 (b). The discrete fields are approximated in the same manner as the finite integration technique. That is, the normal electric field (and flux) is assumed to be constant over the secondary grid face. The tangential magnetic fields are assumed to be constant over the secondary grid edges. Then, employing the finite integration technique of Section 3.4, one can derive the update expression:

$$
E_z^{n+\frac{1}{2}}{}_{i,j,k+\frac{1}{2}} = E_z^{n-\frac{1}{2}}{}_{i,j,k+\frac{1}{2}} +
$$

$$
\frac{\Delta t}{\varepsilon_{i,j,k+\frac{1}{2}}} \left[\left(\frac{H_y^n{}_{i+\frac{1}{2},j,k+\frac{1}{2}} - H_y^n{}_{i-\frac{1}{2},j,k+\frac{1}{2}}}{\ell_{x_i}} \right) - \left(\frac{H_x^n{}_{i,j+\frac{1}{2},k+\frac{1}{2}} - H_x^n{}_{i,j-\frac{1}{2},k+\frac{1}{2}}}{\ell_{y_j}} \right) \right]. \quad (3.116)
$$

Again, an inhomogeneous media is assumed for generality. In this case, following the method of Section 3.7, a weighted average of the dielectric constant arises, such that:

$$
\varepsilon_{i,j,k+\frac{1}{2}} = \frac{(\varepsilon_{i-1,j-1,k+\frac{1}{2}}\Delta x_{i-1} + \varepsilon_{i,j-1,k+\frac{1}{2}}\Delta x_i)\Delta y_{j-1} + (\varepsilon_{i-1,j,k+\frac{1}{2}}\Delta x_{i-1} + \varepsilon_{i,j,k+\frac{1}{2}}\Delta x_i)\Delta y_j}{4\ell_{x_i}\ell_{y_j}}.
$$
$$
(3.117)
$$

The explicit update expressions for the magnetic fields (i.e., (3.115)) is second-order accurate since it is derived via a central difference expression. However, the explicit update expression for the electric field in (3.116) is apparently only first-order accurate since due to non-uniform spacing the discretely sampled electric field is not centered on the secondary grid face. Furthermore, the magnetic field vectors are not necessarily sampled at the secondary grid edge centers. Nevertheless, it can be shown that the global convergence rate of this updating scheme is actually second-order accurate [16, 17].

The non-uniform grid algorithm is stable within the CFL limit providing the time-step satisfies:

$$
\Delta t < \min \left(\frac{\sqrt{\varepsilon_{i,j,k}\mu_{i,j,k}}}{\sqrt{\frac{1}{\Delta x_i^2} + \frac{1}{\Delta y_j^2} + \frac{1}{\Delta z_k^2}}} \right). \quad (3.118)
$$

This implies that the time-step is typically bound by the smallest grid cell in the mesh. However, in practice, a non-uniform grid relaxes the global spatial sampling of a uniform mesh. Consequently, for the same level of accuracy, the non-uniform mesh can reduce the overall computational time and memory quite significantly, and thus it is a very viable approach.

It is noted that when modeling boundaries that do not conform to a rectangular grid (for example skewed boundaries or curved surfaces), the physical boundary must be represented by a "staircase" approximation using the rectangular grid. This leads to some discretization error in the solution. While this is an inherent problem of a rectangular grid, techniques to locally conform the discretization to more arbitrary boundaries are discussed in Chapter 7.

3.11 PROBLEMS

1. Using the secondary grid cell in Fig. 3.2, derive the update expressions (3.29) and (3.30) from Ampère's law.

2. Using the finite integration algorithm described in Section 3.4, derive the electric-field update equations from Ampère's law.

3. Assume a uniform grid that consists of a single primary grid cell, as illustrated in Fig. 3.1. Assume all 6 faces of the grid are perfectly magnetic conducting (PMC) surfaces. (a) Write down the 6 explicit update equations for the H-fields. (b) Write down the 12 explicit update equations for the E-field. To properly write down these equations, apply image theory for the magnetic fields (e.g., (3.92)–(3.95)). (c) Write down in closed form the matrices $\mathbf{C}_e$ and $\mathbf{C}_h$ from parts (a) and (b). (d) Derive a closed form expression for the eigenvalues of $\mathbf{M} = \mathbf{C}_h\mathbf{C}_e$. This can be done using symbolic analysis via a tool such as *Mathematica*™, *Mathcad*™, or *Matlab*™. Show that the maximum eigenvalue is $\lambda_m = 4\left(\frac{1}{\Delta x^2} + \frac{1}{\Delta y^2} + \frac{1}{\Delta z^2}\right)$, and that the eigenspectrum satisfies (3.58).

4. Assuming a uniform staggered grid with cell spacing $\Delta x = \Delta y = \Delta z = \Delta$, compute the *maximum* error in the phase velocity as a function of Δ as Δ is decreased from 0.1λ to 0.001λ. Plot the error in the phase velocity on a log-log scale. Measure the slope and verify second-order convergence. Assume CFLN = 0.99. Repeat for CFLN = 0.5.

5. Repeat Problem 5 for the group velocity.

6. Given the uniform grid spacing $\Delta x = \Delta y = \Delta z = \Delta$, find the maximum grid spacing Δ (in a fraction of a wavelength) such that less than 1 degree of phase error will be expected for a signal propagating over a distance of 1 wavelength. What is Δ that will ensure a phase error of less than 1 degree for a signal propagating over a distance of 10 wavelengths?

7. Prove (3.90).

8. Referring to Fig. 3.9, show that for an inhomogeneous conductivity, that the effective conductivity associated with a primary grid cell (E-field edge) is the average of the conductivities of the four primary cell volumes sharing the edge.

9. Derive the update expression (3.116) using Fig. 3.12 (b). Derive similar update expressions for E_x and E_y for a non-uniform grid.

10. Prove (3.117).

11. Assume a three-dimensional rectangular cavity bound by PEC walls. The cavity has dimensions of 0.5 m × 1 m × 0.25 m, along the x, y, and z-directions, respectively. You are to excite the $TE_{1,m,0}$ modes in this cavity using a z-directed current sheet that is placed in a y-normal

plane, and simulate the cavity resonance using the FDTD method. (a) Derive the z-directed current sheet that will excite the $TE_{1,m,0}$ modes. What is the resonant frequency of the lowest order mode (TE sub 1,1,0). (b) Write a computer program to simulate the fields in the cavity using the FDTD method (c.f., Appendix B). Choose a discretization of $10 \times 20 \times 5$ for this simulation, and place the current sheet appropriately such that the first 4 modes ($m = 1,2,3,4$) will be excited. Assume a time-signature for the current sheet that is a modulated Gaussian pulse. Choose the modulating frequency and bandwidth of the modulated Gaussian pulse such that the first four resonant modes will be excited. (c) Simulate the cavity fields using the FDTD method. Using the FFT, compute the Fourier spectrum of the signal and compare the resonant frequencies with the exact values expected.

REFERENCES

[1] C. R. Paul, K. W. Whites, and S. A. Nasar, *Introduction to Electromagnetic Fields*, 3rd ed. New York: McGraw Hill, 1997. 41, 54

[2] J. A. Stratton, *Electromagnetic Theory*. New York: McGraw-Hill, 1941. 41

[3] K. S. Yee, "Numerical solution of initial boundary value problems involving Maxwell's equations in isotropic media," *IEEE Transactions on Antennas and Propagation*, vol. AP14, pp. 302–307, 1966. DOI: 10.1109/TAP.1966.1138693 41

[4] A. Taflove and M. E. Brodwin, "Numerical-solution of steady-state electromagnetic scattering problems using time-dependent Maxwell's equations," *IEEE Transactions on Microwave Theory and Techniques*, vol. 23, pp. 623–630, 1975. DOI: 10.1109/TMTT.1975.1128640 46, 51

[5] T. Weiland, "Discretization method for solution of Maxwell's equations for 6-component fields," *AEU-International Journal of Electronics and Communications*, vol. 31, pp. 116–120, 1977. xi, 47

[6] R. Courant, K. O. Friedrichs, and H. Lewy, "On the Partial Differential Equations of Mathematical Physics (translated)," *IBM Journal of Research and Development*, vol. 11, pp. 215–234, 1967. DOI: 10.1147/rd.112.0215 48

[7] S. D. Gedney and J. A. Roden, "Numerical stability of nonorthogonal FDTD methods," *IEEE Transactions on Antennas and Propagation*, vol. 48, pp. 231–239, 2000. DOI: 10.1109/8.833072 49

[8] M. Marcus and H. Minc, *Introduction to Linear Algebra*. New York: Dover, 1988. 50

[9] W. H. Press, B. P. Flannery, S. A. Teukolsky, and W. T. Vetterling, *Numerical Recipe's: The Art of Scientific Computing*, 2nd ed. New York: Cambridge University Press, 1992. 54

[10] F. L. Teixeira, "Time-domain finite-difference and finite-element methods for Maxwell equations in complex media," *IEEE Transactions on Antennas and Propagation,* vol. 56, pp. 2150–2166, Aug 2008. DOI: 10.1109/TAP.2008.926767 65

[11] D. Jiao and J. M. Jin, "Time-domain finite-element modeling of dispersive media," *IEEE Microwave And Wireless Components Letters,* vol. 11, pp. 220–222, May 2001. DOI: 10.1109/7260.923034

[12] A. Taflove and S. B. Hagness, *Computational Electrodynamics: The Finite-Difference Time-Domain,* 3rd ed. Boston, MA: Artech House, 2005. 65

[13] R. Luebbers, F. P. Hunsberger, K. S. Kunz, R. B. Standler, and M. Schneider, "A Frequency-Dependent Finite-Difference Time-Domain Formulation for Dispersive Materials," *IEEE Transactions on Electromagnetic Compatibility,* vol. 32, pp. 222–227, 1990. DOI: 10.1109/15.57116 65

[14] R. J. Luebbers and F. Hunsberger, "FDTD for Nth-Order Dispersive Media," *IEEE Transactions on Antennas and Propagation,* vol. 40, pp. 1297–1301, 1992. DOI: 10.1109/8.202707 65

[15] R. M. Joseph, S. C. Hagness, and A. Taflove, "Direct time integration of Maxwell equations in linear dispersive media with absorption for scattering and propagation of femtosecond electromagnetic pulses," *Optics Letters,* vol. 16, pp. 1412–1414, Sep 1991. DOI: 10.1364/OL.16.001412 65

[16] P. Monk and E. Suli, "A convergence analysis of Yee's scheme on non-uniform grids," *Siam Journal on Numerical Analysis,* vol. 31, pp. 393–412, 1994. DOI: 10.1137/0731021 70

[17] P. Monk, "Error estimates for Yee's method on non-uniform grids," *IEEE Transactions on Magnetics,* vol. 30, pp. 3200–3203, 1994. DOI: 10.1109/20.312618 70

CHAPTER 4

Source Excitations

4.1 INTRODUCTION

The purpose of the FDTD method is to provide a solution of Maxwell's equations for real engineering and physics problems. Thus, one must be able to model real physical sources within the discrete FDTD space. For example, if one is modeling a patch antenna excited by a coaxial line feed, then one would have to accurately model the coaxial feed in order to compute the radiation pattern or the input impedance. The role of this chapter is to introduce a number of sources that approximate real physical sources, including: impressed current densities, discrete voltage and current sources, lumped circuit sources, transmission-line excitations, modal excitations, and plane-wave sources. While this is not an exhaustive list of sources, it should serve as an excellent starting point for the analysis of a wide range of engineering problems.

4.2 SOURCE SIGNATURES

The FDTD method simulates *time-dependent* electromagnetic fields. The source excitations will be transient, and when choosing the type of the source, the characteristics of its time signature must be considered. The first consideration is the source bandwidth, which dictates the frequency band over which the electromagnetic fields will be propagated in the FDTD simulation. As seen in the previous chapter, the smallest wavelength is governed by the highest frequency of the band, and thus dictates the overall spatial discretization. On the other hand, the problem being analyzed may also restrict the lower frequency of the band. For example, if one is modeling the excitation of a bounded rectangular waveguide, then one may not want to excite a signal near DC, since such a signal cannot not propagate. Furthermore, for some simulations, it is best to excite a signal with a finite bandwidth about a center frequency. Various signatures with these types of properties are explored in this section.

4.2.1 THE GAUSSIAN PULSE

One of the most commonly used signatures for FDTD source excitations is the Gaussian-pulse:

$$g(t) = e^{-\frac{(t-t_0)^2}{t_w^2}} \qquad (4.1)$$

where t_o is the "time-delay," and t_w is the half width of the pulse. The Gaussian pulse has the characteristic that it is both time-limited and band-limited. The Fourier transform of the Gaussian pulse is:

$$G(f) = t_w \sqrt{\pi} e^{-(\pi f)^2 t_w^2} e^{-j2\pi f t_o}. \tag{4.2}$$

The maximum energy in the pulse is at $f = 0$ Hz. The pulse has a half-bandwidth of $f_{bw} = 1/\pi t_w$ (Hz). The amplitude roles off quickly so that the magnitude drops off to less than 1% at roughly twice the bandwidth. Consequently, the highest frequency of the Gaussian pulse excitation is commonly assumed to be $f_{high} = 2/\pi t_w$. The corresponding wavelength can be used to determine the spatial FDTD discretization.

At time $t = 0$, when the pulse initially turns on, $g(t) = e^{-(t_o/t_w)^2}$, and is not exactly zero. This apparent discontinuity can cause some numerical noise in the simulation. Typically, one should choose $t_o \geq 4t_w$ to minimize high-frequency noise.

4.2.2 THE BLACKMAN-HARRIS WINDOW

An alternative to the Gaussian pulse is the Blackman-Harris window [1, 2]. This function can be written as:

$$b(t) = \begin{cases} \sum_{n=0}^{3} a_n \cos\left(\frac{2\pi n t}{T}\right), & 0 < t < T \\ 0, & \text{otherwise} \end{cases} \tag{4.3}$$

where [2]

$$a = \begin{pmatrix} 0.353222222 \\ -0.488 \\ 0.145 \\ -0.010222222 \end{pmatrix}, \tag{4.4}$$

$$T = \frac{1.55}{f_{bw}}, \tag{4.5}$$

and f_{bw} is the half-bandwidth of the pulse. Similar to the Gaussian pulse, the Blackman-Harris window is both time-limited and frequency-limited. For example, Fig. 4.1 illustrates an overlay of a Gaussian pulse (with $t_o = 4t_w$) and Blackman-Harris window, both with $f_{bw} = 300$ MHz. The pulses have very similar shapes and widths. The distinction is that with the Blackman-Harris window a time-delay is not needed. This reduces the time-delay of turning on the source. The Fourier spectrum of the two pulses is also overlaid in Fig. 4.2. Both have the same cut-off frequency and similar behavior at low frequency. The distinction is that the Blackman-Harris window has a sharper roll-off just past cutoff, but then it decays more slowly at very high frequencies. In both cases, the frequency bandwidth is inversely proportional to the pulse width. Thus, narrower pulses will be broader band- and wider pulses will be narrower banded.

4.2.3 THE DIFFERENTIATED PULSE

The Gaussian and Blackman-Harris pulses may be quite undesirable for some simulations where the source model should have zero energy at DC. One way to mitigate this is to use the derivative

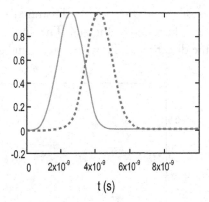

Figure 4.1: Time-domain response of a Blackman-Harris window (solid line) and Gaussian pulse (dashed line) with $f_{bw} = 300$ MHz and $t_o = 4t_w$.

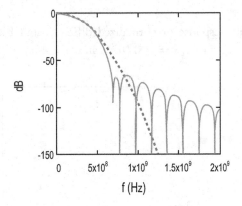

Figure 4.2: Frequency-domain response of a Blackman-Harris window (solid line) and Gaussian pulse (dashed line) with $f_{bw} = 300$ MHz.

of the Gaussian pulse or Blackman-Harris window. The normalized differentiated Gaussian pulse is:

$$d_g(t) = t_w g(t)' = -\frac{(t - t_o)}{t_w} e^{-\frac{(t-t_o)^2}{t_w^2}}. \tag{4.6}$$

Similarly, the normalized Blackman-Harris window is:

$$d_b(t) = \frac{T}{2\pi} b(t)' = \begin{cases} -\sum_{n=0}^{3} a_n n \sin\left(\frac{2\pi n t}{T}\right), & 0 < t < T \\ 0, & \text{otherwise} \end{cases}. \tag{4.7}$$

The time-domain signatures of the two pulses are illustrated in Fig. 4.3. The frequency-domain response of the signals are also illustrated in Fig. 4.4. The upper frequency-band characteristics are assumed to be the same as the direct pulse. However, the signal goes to zero at zero frequency.

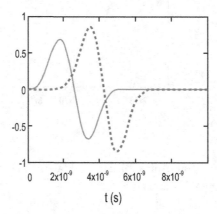

Figure 4.3: Time-domain response of normalized differentiated Blackman-Harries (solid line) and Gaussian pulse (dashed line) with $f_{bw} = 300$ MHz and $t_o = 4t_w$.

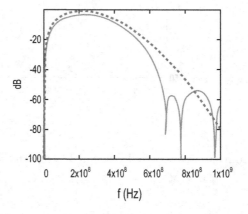

Figure 4.4: Frequency-domain response of normalized differentiated Blackman-Harris (solid line) and Gaussian pulse (dashed line) with $f_{bw} = 300$ MHz.

4.2.4   THE MODULATED PULSE

When simulating radiating structures, such as antennas, or non-TEM waveguide-mode excitations, an excitation that is a time-limited signal with a finite-bandwidth about a center frequency is often required. An effective means of realizing such a signal is to amplitude-modulate the Gaussian pulse, or the Blackman-Harris window. Let f_m be the modulating frequency. The modulated Gaussian pulse is expressed as:

$$m(t) = g(t) \sin (2\pi f_m t).$$

(4.8)

The Blackman-Harris window can also be used. The time-domain response of the signals with both Blackman-Harris and Gaussian pulses is illustrated in Fig. 4.5 for a modulating frequency of 1 GHz and a half bandwidth of 300 MHz. It is noted that the modulated sinusoid is time-limited. The Fourier spectrum of the signals are illustrated in Fig. 4.6. Here it is observed that the signal is indeed centered at the modulating frequency. The upper frequency limit for this excitation that constrains the spatial discretization is $f_{high} = f_m + 2 f_{bw}$.

4.2.5   SINUSOIDAL STEADY-STATE

Certain problems demand a transient simulation with a sinusoidal steady-state source. An example would be the analysis of a non-linear amplifier with a sinusoidal excitation. The sinusoidal source cannot be turned on instantaneously since the slope discontinuity would lead to an unbounded frequency spectrum, which cannot be appropriately modeled within the discrete FDTD algorithm. Rather, the sinusoidal signal must be turned on with a smooth transition. This can be done with either a half Gaussian pulse,

$$s_g(t) = \begin{cases} 0, & t < 0 \\ g(t) \sin (2\pi f_s t), & 0 < t < t_o \\ \sin(2\pi f_s t), & t > t_o \end{cases},$$

(4.9)

or half of a Blackman-Harris window,

$$s_b(t) = \begin{cases} 0, & t < 0 \\ b(t) \sin (2\pi f_s t) / b\left(\frac{T}{2}\right), & 0 < t < \frac{T}{2} \\ \sin(2\pi f_s t), & t > \frac{T}{2} \end{cases}$$

(4.10)

where f_s is the frequency of the sinusoidal source. This leads to a smooth transition to "on" for the transient source, thereby, reducing the bandwidth of the excitation about the steady-state signal. An example of this is illustrated in Fig. 4.7 for a 1 GHz sinusoidal excitation transitioned on with half of a Blackman-Harris window.

When simulating a sinusoidal steady-state signal via the FDTD method, the minimum wavelength is computed based on the frequency $f_{high} = f_s$.

4.3   CURRENT SOURCE EXCITATIONS

4.3.1   VOLUME CURRENT DENSITY

The first-order difference equations representing the FDTD algorithm for lossless, linear, isotropic media are presented in (3.25)–(3.30). The equations can be excited via a volume current density. For

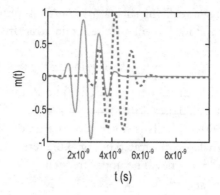

Figure 4.5: Time-domain response of a modulated Gaussian pulse with and $t_o = 4t_w$, $f_m = 1$ GHz, and $f_{bw} = 300$ MHz.

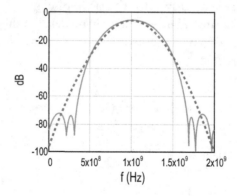

Figure 4.6: Frequency-domain response of modulated Blackman-Harris (solid line) and Gaussian pulse (dashed line) with $t_o = 4t_w$, $f_m = 1$ GHz, and $f_{bw} = 300$ MHz.

example, consider a z-directed volume current density situated over a finite volume. From (3.30), this would contribute to the electric field update:

$$
E_z^{n+\frac{1}{2}}\Big|_{i,j,k+\frac{1}{2}} =
$$

$$
E_z^{n-\frac{1}{2}}\Big|_{i,j,k+\frac{1}{2}} + \frac{\Delta t}{\varepsilon}\left[\left(\frac{H_y^n\big|_{i+\frac{1}{2},j,k+\frac{1}{2}} - H_y^n\big|_{i-\frac{1}{2},j,k+\frac{1}{2}}}{\Delta x}\right) - \left(\frac{H_x^n\big|_{i,j+\frac{1}{2},k+\frac{1}{2}} - H_x^n\big|_{i,j-\frac{1}{2},k+\frac{1}{2}}}{\Delta y}\right) - J_z^n\big|_{i,j,k+\frac{1}{2}}\right]
$$

$$
\tag{4.11}
$$

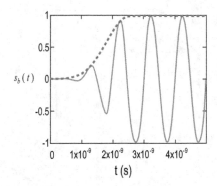

Figure 4.7: Sinusoidal steady-state signal smoothly turned on with half of a Blackman-Harris window, with $f_c = 1$ GHz and $f_{bw} = 300$ MHz.

where $J_z^n \big|_{i,j,k+\frac{1}{2}}$ is the electric current density evaluated at time $t = n\Delta t$. The current density would have a time signature similar to that described in Section 4.2. Each discrete current density occupies a volumetric cell with cross-sectional area $\Delta x \Delta y$ that is centered about the primary grid edge with center at discrete location $(i, j, k + \frac{1}{2})$. The physical current density would then be approximated via a cluster of such cells. In a dual manner, the magnetic current density would contribute to the magnetic field updates.

4.3.2 SURFACE CURRENT DENSITY

A surface current density can be excited on a boundary. For example, an aperture field on a PEC boundary can be approximated via a magnetic current density radiating on the PEC surface. To this end, the surface current density is represented via the boundary condition:

$$\hat{n} \times \vec{E} \big|_S = -\vec{M}_s \tag{4.12}$$

where $\vec{M}_s$ is the surface magnetic current density (V/m). For example, assume that the source is on the $z = 0$ plane. The surface magnetic current density is x-directed. Then:

$$\hat{z} \times \vec{E} \big|_S = -\hat{x} M_{s_x} \Rightarrow E_y \big|_S = M_{s_x}. \tag{4.13}$$

In the discrete space, this would be enforced as a Dirichlet boundary condition:

$$E_y^{n+\frac{1}{2}} \big|_{i,j+\frac{1}{2},0} = M_{s_x}^{n+\frac{1}{2}} \big|_{i,j+\frac{1}{2},0}. \tag{4.14}$$

Note that the x-directed magnetic current would be computed at the spatial position $(i \Delta x, (j + \frac{1}{2})\Delta y, 0)$ and at time $t = (n + \frac{1}{2})\Delta t$, and can have a time-signature similar to that posed in Section 4.2. The surface current density would cover an effective area of dimension $\Delta x \Delta y$ that is

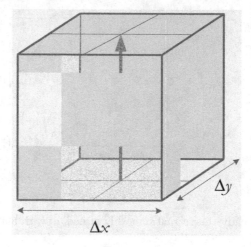

Figure 4.8: Spatial region occupied by a discrete volume current density $J_{z_{i,j,k+\frac{1}{2}}}^{n}$.

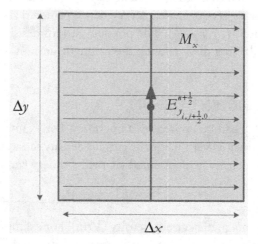

Figure 4.9: Discrete surface patch supporting a surface magnetic current density in the $z = 0$ plane.

centered about the y-directed edge with center $(i\Delta x, (j + \frac{1}{2})\Delta y, 0)$. The physical current density would then be composed of a superposition of such surface patches.

4.4   LUMPED CIRCUIT SOURCE EXCITATIONS

4.4.1   DISCRETE VOLTAGE SOURCE

Many applications require voltage source excitations. Some examples would include printed circuit devices or an antenna excited by a voltage feed. The question is how to accurately represent a voltage source when discretizing the electric and magnetic fields directly. Typically, the physical size of the lumped voltage source is small relative to the wavelength. In such a circumstance, the voltage source can be approximated via a quasi-static representation. That is, the local electric field excited by the voltage source is assumed to be conservative and satisfies the relation:

$$V_s = -\int_a^b \vec{E} \cdot d\vec{\ell}. \tag{4.15}$$

where the line integral is performed over a path between the two nodes (a and b) bounding the lumped voltage source, and V_s is the source voltage.

As an example, consider the voltage source placed between two conducting planes, located in the $k = k_1$ and k_2 planes, respectively, and fixed at discrete (x, y) coordinates (i, j). The voltage source is effectively distributed over the z-directed edges of the primary grid at (i, j) and bound between k_1 and k_2. The source voltage then satisfies the relationship:

$$V_s^{n+\frac{1}{2}} = -\int_{z=k_1 \Delta z}^{k_2 \Delta z} \vec{E}^{n+\frac{1}{2}} \cdot d\vec{\ell} \approx \sum_{k=k_1}^{k_2-1} E_z^{n+\frac{1}{2}}\Big|_{i,j,k+\frac{1}{2}} \Delta z. \tag{4.16}$$

The electric field intensity is assumed to be uniformly distributed along the source. As a consequence, it is approximated as:

$$E_z^{n+\frac{1}{2}}\Big|_{i,j,k+\frac{1}{2}} = -\frac{V_s^{n+\frac{1}{2}}}{\Delta z(k_2 - k_1)}. \tag{4.17}$$

Hard Source: There are two ways in which the voltage source are commonly implemented. The first is what is referred to as a "hard source." The hard source is a boundary condition in which the total electric field is strictly constrained by the voltage source. That is, the FDTD update Equation (3.30) is replaced with (4.17). This is representative of an ideal voltage source with zero internal impedance. Thus, any waves incident upon this source will encounter a short circuit load impedance. Such a source should be used when an actual voltage source with zero internal resistance is desired.

Soft Source: The voltage source can also be implemented as a "soft source." As a soft source, the voltage source is assumed to excite an incident field, which is due to the source alone. The total electric field is thus a superposition of the incident field with the scattered field (which is due to the reaction of the incident field with the device under test). Assume the incident field is then described by (4.17).

The total electric field can then be expressed as a superposition of (4.17) and (3.30), leading to:

$$E_z^{n+\frac{1}{2}}{}_{i,j,k+\frac{1}{2}} = E_z^{n-\frac{1}{2}}{}_{i,j,k+\frac{1}{2}} +$$

$$\frac{\Delta t}{\varepsilon} \left[\left(\frac{H_y^n{}_{i+\frac{1}{2},j,k+\frac{1}{2}} - H_y^n{}_{i-\frac{1}{2},j,k+\frac{1}{2}}}{\Delta x} \right) - \left(\frac{H_x^n{}_{i,j+\frac{1}{2},k+\frac{1}{2}} - H_x^n{}_{i,j-\frac{1}{2},k+\frac{1}{2}}}{\Delta y} \right) \right] - \frac{V_s^{n+\frac{1}{2}}}{\Delta z(k_2 - k_1)}. \tag{4.18}$$

This source will excite an effective voltage between the plates. The difference is that the source will have no load on the wave guiding structure since it is effectively an open circuit impedance. Consequently, any waves reflected back to the source will not be scattered by the source.

Comparing (4.18) and (3.30), it is seen that the soft voltage source can be expressed as an equivalent current source, where the effective current density that excites the desired incident voltage:

$$J_z^n{}_{i,j,k+\frac{1}{2}} = \frac{1}{\eta} \frac{V_s^n}{c\,\Delta t\,\Delta z(k_2 - k_1)}, \tag{4.19}$$

where $\eta = \sqrt{\mu/\varepsilon}$ is the local characteristic wave impedance, and $c = 1/\sqrt{\mu\varepsilon}$ is the speed of light.

4.4.2 DISCRETE THÉVENIN SOURCE

When modeling structures excited by real physical sources, it is often necessary to use a Thévenin equivalent source that contains an internal resistance. Consider the Thévenin source in Fig. 4.10 with open circuit voltage V_{oc} and Thévenin resistance R_{TH}. The source is bound by the nodes n_1

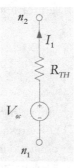

Figure 4.10: A Thévenin equivalent voltage source.

and n_2. Let I_1 represent the branch current through the Thévenin source. The voltage drop across nodes n_1 and n_2 is computed via Kirchoff's voltage law as:

$$V_{1,2} = I_1 R_{TH} - V_{oc}. \tag{4.20}$$

Solving for the branch current:

$$I_1 = \left(V_{1,2} + V_{oc}\right)/R_{TH}. \tag{4.21}$$

The Thévenin source is coupled back into Maxwell's equations through Ampère's law, which is conveniently expressed in the integral form about a secondary grid face as:

$$\frac{\partial}{\partial t}\varepsilon \iint\limits_S \vec{E}\cdot d\vec{s} = \oint\limits_C \vec{H}\cdot d\vec{\ell} - \iint\limits_S \vec{J}\cdot d\vec{s} \tag{4.22}$$

where $\vec{J}$ is assumed to be an impressed current density. Assuming the source is placed on a primary grid edge, then the Thévenin source is bisected by this secondary grid face. Hence,

$$I_1 = \iint\limits_S \vec{J}\cdot d\vec{s} \tag{4.23}$$

is the net current passing through the face.

The voltage drop is related to the electric field using a quasi-static approximation similar to (4.17). For sake of simplicity, assume that the Thévenin source is located on a z-directed, primary grid edge with cell center $(i, j, k + \frac{1}{2})$, and the current in Fig. 4.10 flowing in the $+z$-direction. Then, the local electric field is approximated as:

$$V_{1,2}^n = \Delta z E_z^n{}_{i,j,k+\frac{1}{2}}. \tag{4.24}$$

Combining (4.24), (4.23), and (4.21), we find that

$$\iint\limits_S \vec{J}\cdot d\vec{s} = I_1 = \left(\Delta z E_z^n{}_{i,j,k+\frac{1}{2}} + V_{oc}\right)/R_{TH}. \tag{4.25}$$

This is inserted back into (4.22). The rest of the operator is approximated using the Finite Integration Technique, as detailed in Section 3.4. Consider the secondary-grid face with a z-directed normal bisected by the Thévenin source. Equation (4.22) is then written in discrete form as:

$$\varepsilon \frac{E_z^{n+\frac{1}{2}}{}_{i,j,k+\frac{1}{2}} - E_z^{n-\frac{1}{2}}{}_{i,j,k+\frac{1}{2}}}{\Delta t}\Delta x \Delta y = $$
$$\left[\left(H_y^n{}_{i+\frac{1}{2},j,k+\frac{1}{2}} - H_y^n{}_{i-\frac{1}{2},j,k+\frac{1}{2}}\right)\Delta y - \left(H_x^n{}_{i,j+\frac{1}{2},k+\frac{1}{2}} - H_x^n{}_{i,j-\frac{1}{2},k+\frac{1}{2}}\right)\Delta x - I_1^n\right]. \tag{4.26}$$

The current I_1 is then replaced with (4.25). However, the electric field at time-step n must be approximated via a linear average of the field at time-steps $n \pm \frac{1}{2}$. Consequently,

$$
I_1^n = \frac{\Delta z}{R_{TH}} \left(\frac{E_z^{n+\frac{1}{2}}\Big|_{i,j,k+\frac{1}{2}} + E_z^{n-\frac{1}{2}}\Big|_{i,j,k+\frac{1}{2}}}{2} \right) + \frac{V_{oc}^n}{R_{TH}}.
\tag{4.27}
$$

This is substituted into (4.26). Then, rearranging terms, this is used to derive the first order difference operator used to update the local electric field:

$$
E_z^{n+\frac{1}{2}}\Big|_{i,j,k+\frac{1}{2}} = \frac{\frac{\varepsilon}{\Delta t} - \frac{\Delta z}{\Delta x \Delta y 2 R_{TH}}}{\frac{\varepsilon}{\Delta t} + \frac{\Delta z}{\Delta x \Delta y 2 R_{TH}}} E_z^{n-\frac{1}{2}}\Big|_{i,j,k+\frac{1}{2}} + \frac{1}{\frac{\varepsilon}{\Delta t} + \frac{\Delta z}{\Delta x \Delta y 2 R_{TH}}} \cdot
$$

$$
\left[\left(\frac{H_y^n\big|_{i+\frac{1}{2},j,k+\frac{1}{2}} - H_y^n\big|_{i-\frac{1}{2},j,k+\frac{1}{2}}}{\Delta x} \right) - \left(\frac{H_x^n\big|_{i,j+\frac{1}{2},k+\frac{1}{2}} - H_x^n\big|_{i,j-\frac{1}{2},k+\frac{1}{2}}}{\Delta y} \right) - \frac{V_{oc}^n\big|_{i,j,k+\frac{1}{2}}}{\Delta x \Delta y R_{TH}} \right].
\tag{4.28}
$$

Similar expressions can be derived for sources placed on an x or y-directed edge. This recursive operator is stable, and it provides a second-order accurate representation of a Thévenin equivalent source within the FDTD algorithm.

If the expression in (4.28) is compared to a lossy medium, as discussed in Section 3.8, then it is observed that the local volume has an effective conductivity. That is, comparing (4.28) with (3.99), the effective conductivity is $\sigma = \Delta z / \Delta x \Delta y R_{TH}$ (S/m). Thus, the lumped resistance is effectively distributed over the local discrete volume as a conductivity. Similarly, comparing (4.28) with (4.11), the voltage source is represented by an equivalent current density, with $J_z^n\big|_{i,j,k+\frac{1}{2}} = V_{oc}^n\big|_{i,j,k+\frac{1}{2}} \Big/ \Delta x \Delta y R_{TH}$.

This is analogous to treating the Thévenin voltage source as an equivalent Norton current source.

It is observed that if the voltage source is forced to zero, then the update equation in (4.28) represents the update equation for an equivalent resistive load R_{TH} placed on the primary grid edge. Thus, this provides the means to add resistive loads to the FDTD method.

The Thévenin source may be distributed over a number of edges. If the sources are in series, then, applying Kirchoff's voltage law, the series voltage sources add, and the series resistances add. In practice, one would evenly distribute R_{TH} and V_{oc} over all the edges in the series.

On the other hand, consider the case when identical Thévenin sources are placed in parallel. Again, applying Kirchoff's laws, the net resistance is the total parallel resistance (which sums inversely). However, the voltage is the same as a single source.

4.4.3   LUMPED LOADS

In the previous section, it was found via the derivation of a lumped voltage source, that one can also model a lumped resistance directly within the FDTD method. This is very useful when modeling circuit devices or antennas. Other lumped element loads can also be useful. Passive loads such as capacitors or inductors can be modeled in a similar manner as the lumped resistance. That is, referring to Fig. 4.10, let I_1 be the current flowing through a lumped circuit element, which for sake of illustration, is a z-directed primary grid edge with grid center $(i, j, k + \frac{1}{2})$. Assume that the lumped circuit element is a capacitor. The lumped circuit voltage-current relationship for the capacitor is:

$$I = C\frac{dV}{dt} \tag{4.29}$$

where C is the branch capacitance in Farads (F). From (4.16), we can let $V^{n+\frac{1}{2}} = \Delta z E_z^{n+\frac{1}{2}}\Big|_{i,j,k+\frac{1}{2}}$. Applying a central difference approximation, (4.29) can be expressed in the discrete FDTD space as:

$$I_1^n = C\frac{\Delta z}{\Delta t}\left(E_z^{n+\frac{1}{2}}\Big|_{i,j,k+\frac{1}{2}} - E_z^{n-\frac{1}{2}}\Big|_{i,j,k+\frac{1}{2}}\right). \tag{4.30}$$

This is plugged into the discrete form of (4.22) expressed in (4.26):

$$\varepsilon\frac{E_z^{n+\frac{1}{2}}\Big|_{i,j,k+\frac{1}{2}} - E_z^{n-\frac{1}{2}}\Big|_{i,j,k+\frac{1}{2}}}{\Delta t}\Delta x\Delta y =$$

$$\left[\left(H_y^n\Big|_{i+\frac{1}{2},j,k+\frac{1}{2}} - H_y^n\Big|_{i-\frac{1}{2},j,k+\frac{1}{2}}\right)\Delta y - \left(H_x^n\Big|_{i,j+\frac{1}{2},k+\frac{1}{2}} - H_x^n\Big|_{i,j-\frac{1}{2},k+\frac{1}{2}}\right)\Delta x\right] -$$

$$C\frac{\Delta z}{\Delta t}\left(E_z^{n+\frac{1}{2}}\Big|_{i,j,k+\frac{1}{2}} - E_z^{n-\frac{1}{2}}\Big|_{i,j,k+\frac{1}{2}}\right) \tag{4.31}$$

This leads to the discrete update equation:

$$E_z^{n+\frac{1}{2}}\Big|_{i,j,k+\frac{1}{2}} = E_z^{n-\frac{1}{2}}\Big|_{i,j,k+\frac{1}{2}} + \frac{\Delta t}{\varepsilon + C\frac{\Delta z}{\Delta x\Delta y}}\left[\left(\frac{H_y^n\Big|_{i+\frac{1}{2},j,k+\frac{1}{2}} - H_y^n\Big|_{i-\frac{1}{2},j,k+\frac{1}{2}}}{\Delta x}\right)\right.$$

$$\left. - \left(\frac{H_x^n\Big|_{i,j+\frac{1}{2},k+\frac{1}{2}} - H_x^n\Big|_{i,j-\frac{1}{2},k+\frac{1}{2}}}{\Delta y}\right)\right]. \tag{4.32}$$

Interestingly, the local lumped capacitance thus appears as a local anisotropy of the dielectric, with local permittivity $\varepsilon_{z,z} = \varepsilon + C\Delta z/\Delta x\Delta y$ (F/m).

It is interesting to observe that one can also achieve a parallel RC load by simply combining the local resistance and capacitance models. Applying Kirchoff's current law, and combining (4.27) and (4.30) the total lumped element branch current is:

$$I_1^n = \frac{\Delta z}{R}\left(\frac{E_z^{n+\frac{1}{2}}_{i,j,k+\frac{1}{2}} + E_z^{n-\frac{1}{2}}_{i,j,k+\frac{1}{2}}}{2}\right) + C\frac{\Delta z}{\Delta t}\left(E_z^{n+\frac{1}{2}}_{i,j,k+\frac{1}{2}} - E_z^{n-\frac{1}{2}}_{i,j,k+\frac{1}{2}}\right). \tag{4.33}$$

Combining this with (4.26), leads to the update equation:

$$E_z^{n+\frac{1}{2}}_{i,j,k+\frac{1}{2}} = \frac{\frac{\varepsilon}{\Delta t} + \frac{C}{\Delta t}\frac{\Delta z}{\Delta x\Delta y} - \frac{\Delta z}{\Delta x\Delta y 2R}}{\frac{\varepsilon}{\Delta t} + \frac{C}{\Delta t}\frac{\Delta z}{\Delta x\Delta y} + \frac{\Delta z}{\Delta x\Delta y 2R}} E_z^{n-\frac{1}{2}}_{i,j,k+\frac{1}{2}} + \frac{1}{\frac{\varepsilon}{\Delta t} + \frac{C}{\Delta t}\frac{\Delta z}{\Delta x\Delta y} + \frac{\Delta z}{\Delta x\Delta y 2R}} \cdot$$

$$\left[\left(\frac{H_y^n_{i+\frac{1}{2},j,k+\frac{1}{2}} - H_y^n_{i-\frac{1}{2},j,k+\frac{1}{2}}}{\Delta x}\right) - \left(\frac{H_x^n_{i,j+\frac{1}{2},k+\frac{1}{2}} - H_x^n_{i,j-\frac{1}{2},k+\frac{1}{2}}}{\Delta y}\right)\right] \tag{4.34}$$

Consider next a lumped inductor load. The voltage-current relationship of a branch inductance is:

$$V = L\frac{dI}{dt}, \tag{4.35}$$

where L is the branch inductance in Henrys (H). Because of the derivative on the current, I is treated as an unknown, which must be updated at each time-step. Applying central differencing:

$$\left(\frac{I_1^n - I_1^{n-1}}{\Delta t}\right) = \frac{1}{L}V^{n+\frac{1}{2}} = \frac{\Delta z}{L}E_z^{n-\frac{1}{2}}_{i,j,k+\frac{1}{2}}. \tag{4.36}$$

Therefore,

$$I_1^n = I_1^{n-1} + \Delta t\frac{\Delta z}{L}E_z^{n-\frac{1}{2}}_{i,j,k+\frac{1}{2}}. \tag{4.37}$$

The update for the electric field is then simply expressed as:

$$E_z^{n+\frac{1}{2}}_{i,j,k+\frac{1}{2}} = E_z^{n-\frac{1}{2}}_{i,j,k+\frac{1}{2}} + \frac{\Delta t}{\varepsilon}\left[\left(\frac{H_y^n_{i+\frac{1}{2},j,k+\frac{1}{2}} - H_y^n_{i-\frac{1}{2},j,k+\frac{1}{2}}}{\Delta x}\right)\right.$$

$$\left.- \left(\frac{H_x^n_{i,j+\frac{1}{2},k+\frac{1}{2}} - H_x^n_{i,j-\frac{1}{2},k+\frac{1}{2}}}{\Delta y}\right) - I_1^n\right], \tag{4.38}$$

where I_1^n is updated by (4.37) at each time-step.

One can also treat a series inductor-resistor lumped circuit. The voltage-current relationship of the branch network is then:

$$V = L\frac{dI}{dt} + IR,\tag{4.39}$$

where V is the voltage across both series elements. This can be approximated in discrete form as:

$$L\left(\frac{I_1^n - I_1^{n-1}}{\Delta t}\right) + R\left(\frac{I_1^n + I_1^{n-1}}{2}\right) = \Delta z E_z^{n-\frac{1}{2}}\Big|_{i,j,k+\frac{1}{2}}.\tag{4.40}$$

This leads to the update equation for the branch current:

$$I_1^n = \frac{\frac{L}{\Delta t} - \frac{R}{2}}{\frac{L}{\Delta t} + \frac{R}{2}} I_1^{n-1} + \frac{\Delta z}{\frac{L}{\Delta t} + \frac{R}{2}} E_z^{n-\frac{1}{2}}\Big|_{i,j,k+\frac{1}{2}}.\tag{4.41}$$

Finally, the field update equation for the circuit loaded edge is (4.38), where I_1^n is updated by (4.41) at each time-step.

In this section, local field updates were derived for specific lumped loads. For more information regarding a more general formulation that can handle arbitrary loads, including multi-port and non-linear devices, the reader is referred to the work by Thomas, et al. [3].

4.5 PLANE WAVE EXCITATION

4.5.1 THE TOTAL-FIELD SCATTERED FIELD FORMULATION

For many practical problems, it is necessary to employ a plane-wave excitation. It is quite impractical to directly model the source that radiates the plane wave since the source generally must extend beyond or even lie completely outside of the FDTD problem domain. Thus, alternate means of exciting the plane wave source must be sought out. Perhaps one of the most expedient methods is to introduce a "plane-wave boundary," or Huygen surface, that injects the plane wave into the problem domain [4]. This method is now referred to as the *total-field/scattered-field* formulation, which will be described in the remainder of this section.

Here it is assumed that the FDTD method is being used to model the scattering of a plane wave off some device under test (DUT). The DUT is assumed to be situated in a homogeneous, isotropic, lossless material, which for simplicity is assumed to be a homogeneous free space. The electric and magnetic fields can be expressed as a superposition of incident and scattered fields:

$$\vec{E}^{tot} = \vec{E}^{inc} + \vec{E}^{scat}, \qquad \vec{H}^{tot} = \vec{H}^{inc} + \vec{H}^{scat}.\tag{4.42}$$

The incident fields ($\vec{E}^{inc}$, $\vec{H}^{inc}$) are the plane wave electromagnetic fields propagating in the homogeneous free space in the *absence* of the DUT. The scattered fields ($\vec{E}^{scat}$, $\vec{H}^{scat}$) are the perturbation of the field due to the presence of the DUT. The total field is the actual physical field.

Let Ω be the global space defining the FDTD discretization. Let Ω^{tot} be the interior region of Ω that completely contains the DUT. Let Ω^{scat} be the region exterior to Ω^{tot}, such that $\Omega = \Omega^{tot} \cup \Omega^{scat}$. The two regions share the boundary $\partial\Omega^{tot/scat}$. This is illustrated in Fig. 4.11. Within Ω^{tot}, the total electric and magnetic fields ($\vec{E}^{tot}$, $\vec{H}^{tot}$) are represented via the FDTD formulation. The scattered fields ($\vec{E}^{scat}$, $\vec{H}^{scat}$) are represented within Ω^{scat}.

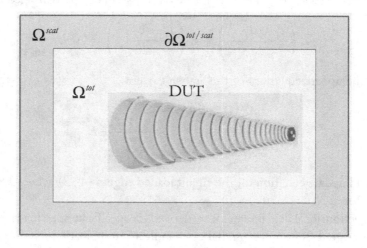

Figure 4.11: FDTD domain broken up into total field and scattered field regions.

The boundary $\partial\Omega^{tot/scat}$ is assumed to be a surface defined of primary grid faces. Typically, this boundary will be rectangular in shape. All fields within Ω^{tot} and on $\partial\Omega^{tot/scat}$ are assumed to be total fields. All fields within Ω^{scat}, excluding $\partial\Omega^{tot/scat}$ are assumed to be scattered fields. Consider the update of the tangential field E_x^{tot} that lies on the plane wave boundary $\partial\Omega^{tot/scat}$, as illustrated in Fig. 4.12. The planar surface of $\partial\Omega^{tot/scat}$ has a $+y$-directed normal into Ω^{scat}. From (3.28), the update for E_x^{tot} is:

$$
E_{x \atop i+\frac{1}{2},j,k}^{tot^{n+\frac{1}{2}}} = E_{x \atop i+\frac{1}{2},j,k}^{tot^{n-\frac{1}{2}}} + \frac{\Delta t}{\varepsilon}\left[\left(\frac{H_{z \atop i+\frac{1}{2},j+\frac{1}{2},k}^{tot^n} - H_{z \atop i+\frac{1}{2},j-\frac{1}{2},k}^{tot^n}}{\Delta y}\right) - \left(\frac{H_{y \atop i+\frac{1}{2},j,k+\frac{1}{2}}^{tot^n} - H_{y \atop i+\frac{1}{2},j,k-\frac{1}{2}}^{tot^n}}{\Delta z}\right)\right].
\tag{4.43}
$$

From Fig. 4.12, it is observed that $H_{z \atop i+\frac{1}{2},j+\frac{1}{2},k}^{tot^n}$ is in the scattered field region Ω^{scat}, where there is not a description of H_z^{tot}. Rather, H_z^{scat} is discretized. To remedy this, H_z^{tot} is expressed in terms of

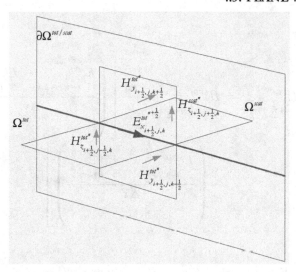

Figure 4.12: Discrete electric field projection tangential to $\delta\Omega^{tot/scat}$ and the magnetic field samples required for the Yee-update of the field.

H_z^{scat} via (4.42). As a consequence, (4.43) becomes:

$$
\begin{aligned}
E_x^{tot^{n+\frac{1}{2}}}_{i+\frac{1}{2},j,k} = E_x^{tot^{n-\frac{1}{2}}}_{i+\frac{1}{2},j,k} + \frac{\Delta t}{\varepsilon} &\left[\left(\frac{H_z^{scat^n}_{i+\frac{1}{2},j+\frac{1}{2},k} - H_z^{tot^n}_{i+\frac{1}{2},j-\frac{1}{2},k}}{\Delta y} \right) \right. \\
&\left. - \left(\frac{H_y^{tot^n}_{i+\frac{1}{2},j,k+\frac{1}{2}} - H_y^{tot^n}_{i+\frac{1}{2},j,k-\frac{1}{2}}}{\Delta z} \right) \right] + \frac{\Delta t}{\varepsilon} \frac{H_z^{inc^n}_{i+\frac{1}{2},j+\frac{1}{2},k}}{\Delta y}
\end{aligned}
\tag{4.44}
$$

where $H_z^{inc^n}_{i+\frac{1}{2},j+\frac{1}{2},k}$ is the actual incident field computed at the primary face cell center at time $t = n\Delta t$. Similar expressions are derived for all tangential electric fields on the plane wave boundary $\partial\Omega^{tot/scat}$.

Next, consider the magnetic field updates. The discrete total magnetic field sampled on the plane wave boundary is normal to the boundary $\partial\Omega^{tot/scat}$. The normal magnetic field updates using (3.25)–(3.27) involve only total electric fields. However, the update of the magnetic field tangential to $\partial\Omega^{tot/scat}$, and one half of a space cell *outside* of $\partial\Omega^{tot/scat}$ (i.e., inside Ω^{scat}) requires the sampling of the total electric field. For example, consider the update of $H_z^{scat^n}_{i+\frac{1}{2},j+\frac{1}{2},k}$ in Fig. 4.12.

From (3.27),

$$
H_{z\,_{i+\frac{1}{2},j+\frac{1}{2},k}}^{scat^{n+1}} = H_{z\,_{i+\frac{1}{2},j+\frac{1}{2},k}}^{scat^{n}} + \frac{\Delta t}{\mu}\left[\left(\frac{E_{x\,_{i+\frac{1}{2},j+1,k}}^{scat^{n+\frac{1}{2}}} - E_{x\,_{i+\frac{1}{2},j,k}}^{scat^{n+\frac{1}{2}}}}{\Delta y}\right)\right.
$$
$$
\left.-\left(\frac{E_{y\,_{i+1,j+\frac{1}{2},k}}^{scat^{n+\frac{1}{2}}} - E_{y\,_{i,j+\frac{1}{2},k}}^{scat^{n+\frac{1}{2}}}}{\Delta x}\right)\right]. \tag{4.45}
$$

It is observed from Fig. 4.12, that $E_{x\,_{i+\frac{1}{2},j,k}}^{n+\frac{1}{2}}$ is on $\partial\Omega^{tot/scat}$, where it is represented by the total field. From (4.42), the scattered field can be expressed as a function of the total field as $E_x^{scat} = E_x^{tot} - E_x^{inc}$. Consequently, (4.45) can be written as:

$$
H_{z\,_{i+\frac{1}{2},j+\frac{1}{2},k}}^{scat^{n+1}} = H_{z\,_{i+\frac{1}{2},j+\frac{1}{2},k}}^{scat^{n}} + \frac{\Delta t}{\mu}\left[\left(\frac{E_{x\,_{i+\frac{1}{2},j+1,k}}^{scat^{n+\frac{1}{2}}} - E_{x\,_{i+\frac{1}{2},j,k}}^{tot^{n+\frac{1}{2}}}}{\Delta y}\right)\right.
$$
$$
\left.-\left(\frac{E_{y\,_{i+1,j+\frac{1}{2},k}}^{scat^{n+\frac{1}{2}}} - E_{y\,_{i,j+\frac{1}{2},k}}^{scat^{n+\frac{1}{2}}}}{\Delta x}\right)\right] + \frac{\Delta t}{\mu}\frac{E_{x\,_{i+\frac{1}{2},j,k}}^{inc^{n+\frac{1}{2}}}}{\Delta y} \tag{4.46}
$$

where $E_{x\,_{i+\frac{1}{2},j,k}}^{inc^{n+\frac{1}{2}}}$ is the x-projection of the incident electric field computed at the discrete coordinate $(i + \frac{1}{2}, j, k)$ at time $t = (n + \frac{1}{2})\Delta t$.

Observing the updates (4.44) and (4.46), it is found that the updates are identical to source-free updates in (3.25)–(3.30) with the additional incident field term superimposed. Thus, in a practical implementation of the method, one would employ the standard updates of (3.25)–(3.30) first, and then superimpose the plane-wave boundary sources afterwards. For example, let the total field region be a rectangular box with bounding coordinates $(i_1^{pw}, j_1^{pw}, k_1^{pw})$ and $(i_2^{pw}, j_2^{pw}, k_2^{pw})$. The fields in the scattered and total field regions would first be updated using (3.25)–(3.30). After this, the incident plane wave fields would be injected using the following procedure:

$$
E_x^{n+\frac{1}{2}}\Big|_{i+\frac{1}{2},j,k} = E_x^{n+\frac{1}{2}}\Big|_{i+\frac{1}{2},j,k} - \frac{\Delta t}{\varepsilon} \frac{H_z^{inc^n}\Big|_{i+\frac{1}{2},j-\frac{1}{2},k}}{\Delta y}\Bigg|_{i=i_1^{pw}..i_2^{pw}-1;\ j=j_1^{pw};\ k=k_1^{pw}..k_2^{pw}} \tag{4.47}
$$

$$
E_x^{n+\frac{1}{2}}\Big|_{i+\frac{1}{2},j,k} = E_x^{n+\frac{1}{2}}\Big|_{i+\frac{1}{2},j,k} + \frac{\Delta t}{\varepsilon} \frac{H_z^{inc^n}\Big|_{i+\frac{1}{2},j+\frac{1}{2},k}}{\Delta y}\Bigg|_{i=i_1^{pw}..i_2^{pw}-1;\ j=j_2^{pw};\ k=k_1^{pw}..k_2^{pw}} \tag{4.48}
$$

$$
E_x^{n+\frac{1}{2}}\Big|_{i+\frac{1}{2},j,k} = E_x^{n+\frac{1}{2}}\Big|_{i+\frac{1}{2},j,k} + \frac{\Delta t}{\varepsilon} \frac{H_y^{inc^n}\Big|_{i+\frac{1}{2},j,k-\frac{1}{2}}}{\Delta y}\Bigg|_{i=i_1^{pw}..i_2^{pw}-1;\ j=j_1^{pw}..j_2^{pw};\ k=k_1^{pw}} \tag{4.49}
$$

$$
E_x^{n+\frac{1}{2}}\Big|_{i+\frac{1}{2},j,k} = E_x^{n+\frac{1}{2}}\Big|_{i+\frac{1}{2},j,k} - \frac{\Delta t}{\varepsilon} \frac{H_y^{inc^n}\Big|_{i+\frac{1}{2},j,k+\frac{1}{2}}}{\Delta y}\Bigg|_{i-i_1^{pw}..i_2^{pw}-1;\ j=j_1^{pw}..j_2^{pw};\ k=k_2^{pw}} . \tag{4.50}
$$

Similar injections are applied to E_y and E_z on the plane wave boundaries.
For the magnetic fields:

$$
H_x^{n+1}\Big|_{i,j-\frac{1}{2},k+\frac{1}{2}} = H_x^{n+1}\Big|_{i,j-\frac{1}{2},k+\frac{1}{2}} + \frac{\Delta t}{\mu} \frac{E_z^{inc^{n+\frac{1}{2}}}\Big|_{i,j,k+\frac{1}{2}}}{\Delta y}\Bigg|_{i=i_1^{pw}..i_2^{pw};\ j=j_1^{pw};\ k=k_1^{pw}..k_2^{pw}-1} \tag{4.51}
$$

$$
H_x^{n+1}\Big|_{i,j+\frac{1}{2},k+\frac{1}{2}} = H_x^{n+1}\Big|_{i,j+\frac{1}{2},k+\frac{1}{2}} - \frac{\Delta t}{\mu} \frac{E_z^{inc^{n+\frac{1}{2}}}\Big|_{i,j,k+\frac{1}{2}}}{\Delta y}\Bigg|_{i=i_1^{pw}..i_2^{pw};\ j=j_2^{pw};\ k=k_1^{pw}..k_2^{pw}-1} \tag{4.52}
$$

$$
H_x^{n+1}\Big|_{i,j+\frac{1}{2},k-\frac{1}{2}} = H_x^{n+1}\Big|_{i,j+\frac{1}{2},k-\frac{1}{2}} - \frac{\Delta t}{\mu} \frac{E_y^{inc^{n+\frac{1}{2}}}\Big|_{i,j+\frac{1}{2},k}}{\Delta z}\Bigg|_{i=i_1^{pw}..i_2^{pw};\ j=j_1^{pw}..j_2^{pw}-1;\ k=k_1^{pw}} \tag{4.53}
$$

$$H_{x}^{n+1}\Big|_{i,j+\frac{1}{2},k+\frac{1}{2}} = H_{x}^{n+1}\Big|_{i,j+\frac{1}{2},k+\frac{1}{2}} + \frac{\Delta t}{\mu} \frac{E_{y}^{inc\,n+\frac{1}{2}}\Big|_{i,j+\frac{1}{2},k}}{\Delta z}\Bigg|_{i=i_{1}^{pw}..i_{2}^{pw};\ j=j_{1}^{pw}..j_{2}^{pw}-1;\ k=k_{2}^{pw}} . \quad (4.54)$$

Similar injections are applied to H_y and H_z one-half of a cell outside of the plane wave boundaries.

4.5.2 GENERAL DESCRIPTION OF A UNIFORM PLANE WAVE

An arbitrary uniform plane wave can be expressed in a spherical coordinate system, as defined by Fig. 4.13. The direction of propagation is assumed to be along the negative radial direction. This is

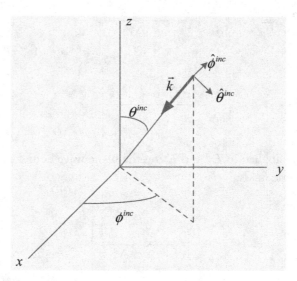

Figure 4.13: Plane wave propagating along the radial propagation vector $\vec{k}$ defined by the spherical coordinates (θ, ϕ) and the local unit spherical unitary vectors.

illustrated by the $\vec{k}$-vector, known as the propagation vector. The plane wave is defined by an angle of incidence of $(\theta^{inc}, \phi^{inc})$, where θ^{inc} is the zenith angle, or the angle between the vertical z-axis and the $\vec{k}$-vector, and ϕ^{inc} is the azimuthal angle, or the angle between the projection of the $\vec{k}$-vector on the $x - y$ plane and the $x-$ axis. The k-vector is defined as a function of $(\theta^{inc}, \phi^{inc})$ as:

$$\vec{k} = -k\left(\sin\theta^{inc}\cos\phi^{inc}\hat{x} + \sin\theta^{inc}\sin\phi^{inc}\hat{y} + \cos\theta^{inc}\hat{z}\right) = -k\hat{\rho} \quad (4.55)$$

where $k = \omega/c = 2\pi/\lambda$ is the wave number, and c is the speed of light.

The arbitrary incident field can be decoupled into a superposition of *vertical* and *horizontal* polarized waves, which are defined by time-harmonic electric fields

$$\vec{E}_V^{inc}\left(\vec{r};\theta^{inc},\phi^{inc}\right) = E_{V_o}e^{-j\vec{k}\cdot\vec{r}}e^{j\omega t}\hat{\theta}^{inc}, \tag{4.56}$$

$$\vec{E}_H^{inc}\left(\vec{r};\theta^{inc},\phi^{inc}\right) = E_{H_o}e^{-j\vec{k}\cdot\vec{r}}e^{j\omega t}\hat{\phi}^{inc}, \tag{4.57}$$

respectively, where,

$$\hat{\theta}^{inc} = \cos\theta^{inc}\cos\phi^{inc}\hat{x} + \cos\theta^{inc}\sin\phi^{inc}\hat{y} - \sin\theta^{inc}\hat{z}, \tag{4.58}$$

$$\hat{\phi} = -\sin\phi^{inc}\hat{x} + \cos\phi^{inc}\hat{y}, \tag{4.59}$$

and

$$\vec{r} = x\hat{x} + y\hat{y} + z\hat{z}. \tag{4.60}$$

The magnetic field is derived from the electric field via Faraday's law, leading to:

$$\vec{H}_V^{inc}\left(\vec{r};\theta^{inc},\phi^{inc}\right) = \frac{-E_{V_o}}{\eta_0}e^{-j\vec{k}\cdot\vec{r}}e^{j\omega t}\hat{\phi}^{inc} \tag{4.61}$$

$$\vec{H}_H^{inc}\left(\vec{r};\theta^{inc},\phi^{inc}\right) = \frac{E_{H_o}}{\eta_0}e^{-j\vec{k}\cdot\vec{r}}e^{j\omega t}\hat{\theta}^{inc} \tag{4.62}$$

where, η_0 is the wave characteristic impedance of the host medium.

Applying Fourier transform theory [5], the time-harmonic plane-waves can be transformed into the time-domain, leading to:

$$\vec{E}_V^{inc}\left(t,\vec{r},\theta^{inc},\phi^{inc}\right) = E_{V_o}f\left(t+\frac{\hat{\rho}\cdot\vec{r}}{c}\right)\hat{\theta}^{inc},\ \vec{H}_V^{inc}\left(t,\vec{r},\theta^{inc},\phi^{inc}\right) = -\frac{E_{V_o}}{\eta_o}f\left(t+\frac{\hat{\rho}\cdot\vec{r}}{c}\right)\hat{\phi}^{inc} \tag{4.63}$$

$$\vec{E}_H^{inc}\left(t,\vec{r},\theta^{inc},\phi^{inc}\right) = E_{H_o}f\left(t+\frac{\hat{\rho}\cdot\vec{r}}{c}\right)\hat{\phi}^{inc},\ \vec{H}_H^{inc}\left(t,\vec{r},\theta^{inc},\phi^{inc}\right) = \frac{E_{H_o}}{\eta_o}f\left(t+\frac{\hat{\rho}\cdot\vec{r}}{c}\right)\hat{\theta}^{inc} \tag{4.64}$$

where, $f\left(t + \hat{\rho}\cdot\vec{r}/c\right)$ represents the time/space-signature of the transient plane wave propagating in the negative radial direction. The single argument function f is appropriately chosen to be one of the different signatures detailed in Section 4.2.

At the initial time $t = 0$, the traveling wave function $f\left(t+\hat{\rho}\cdot\vec{r}/c\right)$ can have a non-zero argument at most spatial locations (that is for all $\hat{\rho}\cdot\vec{r} > 0$). However, the incident field on the plane wave boundary should initially be zero. To circumvent this, the spatial origin is shifted to a point $\vec{r}_o$ outside of the plane wave boundary, such that the argument $\hat{\rho}\cdot(\vec{r}-\vec{r}_o)/c < 0$ at all coordinates on $\partial\Omega^{tot/scat}$. Let $\vec{r}_{\min}$ and $\vec{r}_{\max}$ represent the minimum and maximum coordinates of $\partial\Omega^{tot/scat}$. We can then define the shifted origin to be

$$\vec{r}_o = x_o\hat{x} + y_o\hat{y} + z_o\hat{z} \tag{4.65}$$

where,

$$x_o = \begin{cases} x_{\max} & \text{if } \frac{-\pi}{2} \leq \phi^{inc} \leq +\frac{\pi}{2} \\ x_{\min} & \text{else} \end{cases} \tag{4.66}$$

$$y_o = \begin{cases} y_{\max} & \text{if } 0 \leq \phi^{inc} \leq \pi \\ y_{\min} & \text{else} \end{cases} \tag{4.67}$$

and

$$z_o = \begin{cases} z_{\max} & \text{if } 0 \leq \theta^{inc} \leq \frac{\pi}{2} \\ z_{\min} & \text{else} \end{cases} . \tag{4.68}$$

The time dependent plane wave $\vec{E}_V^{inc}$ with the shifted origin is

$$\vec{E}_V^{inc}\left(t, \vec{r}, \theta^{inc}, \phi^{inc}\right) = E_{V_o} f\left(t + \frac{\hat{\rho} \cdot (\vec{r} - \vec{r}_o)}{c}\right) \hat{\theta}^{inc}. \tag{4.69}$$

Similar expressions are derived for $\vec{H}_V^{inc}$, $\vec{E}_H^{inc}$, and $\vec{H}_H^{inc}$ from (4.63) and (4.64).

4.5.3   COMPUTING THE DISCRETE INCIDENT FIELD VECTOR

The time-dependent plane waves were detailed in the previous section. The Cartesian projections of the vertically or horizontally polarized plane waves are computed via (4.69), (4.63), (4.64), (4.58), and (4.59). These functions can be evaluated analytically by first computing the argument $\xi = t + \hat{\rho} \cdot (\vec{r} - \vec{r}_o)/c$ at the discrete coordinate, and then evaluate the signature function $f(\xi)$. For large three-dimensional problems, the cost of this calculation at each time-step can become quite expensive. Rather this can be computed more efficiently in the following manner.

It is realized that the spatial dependence of the plane-wave argument $\hat{\rho} \cdot (\vec{r} - \vec{r}_o)/c$ is constant with respect to time for each grid coordinate. Consequently, this can be pre-computed prior to the time-dependent simulation. Next, since the signature function $f(\xi)$ is simply a one-dimensional function, it can be pre-computed and stored in an array. The discrete sampling of the function can conveniently be chosen to be the time-discretization Δt. Consequently, we can define

$$f^n = f(n\Delta t) \tag{4.70}$$

where $n > 0$.

Next, define the discrete value for the spatial portion of the argument

$$\zeta_{i,j,k} = \left(\frac{\hat{\rho} \cdot (\vec{r}_{i,j,k} - \vec{r}_o)/c}{\Delta t}\right) \tag{4.71}$$

where $\vec{r}_{i,j,k}$ is the discrete FDTD grid coordinate of the incident electric field's primary grid edge center or the magnetic field's primary grid face center. Next, the discrete integer value is defined

$$m_{i,j,k} = \text{floor}(\zeta_{i,j,k}) \tag{4.72}$$

where "floor" is a function that truncates the argument to the nearest integer. For example, floor(3.71) = 3. For negative numbers, floor(-3.71) = -4. $\zeta_{i,j,k}$ and $m_{i,j,k}$ can be pre-computed and stored *a priori* to the FDTD simulation for each grid edge or face center, for which the incident field will need to be calculated. Then, during the time-domain simulation, the time signature at coordinate $\vec{r}_{i,j,k}$ is approximated via the function:

$$f\left(n\Delta t + \frac{\hat{\rho}\cdot(\vec{r}_{i,j,k}-\vec{r}_o)}{c}\right) \approx \begin{cases} 0, & \text{if } n+m_{i,j,k} < 0 \\ (\zeta_{i,j,k}-m_{i,j,k})\left(f^{m_{i,j,k}+n+1}-f^{m_{i,j,k}+n}\right)+f^{m_{i,j,k}+n}, & \text{else.} \end{cases}$$
(4.73)

Note that for the electric field updates, as defined in (4.47)–(4.50), the argument will be offset for the time-step $n+1/2$. This is expediently done by storing a second one-dimensional array:

$$f^{n+\frac{1}{2}} = f\left(\left(n+\tfrac{1}{2}\right)\Delta t\right)$$
(4.74)

and replacing n with $n+1/2$ in (4.73).

4.5.4 NUMERICAL DISPERSION

The strategy presented in this section is to propagate a plane wave through the FDTD grid. In fact, in the absence of the DUT, the plane wave will launch from the plane-wave boundary near $\vec{r}_o$, and will propagate through the total field region, and then most remarkably be absorbed by the plane-wave boundary on the other side. This happens because the incident field superimposed at the boundary is actually cancelling the incident field propagating in the volume. As we learned in Chapter 3, the wave propagating in the discrete FDTD space propagates with a wave vector that suffers from numerical dispersion, which is also anisotropic. However, the analytical form used to compute the propagation of the incident wave assumed the exact physical wave-number. As a consequence, the wave that propagates through the discrete FDTD grid will not exactly cancel with the incident field computed via an analytical expression. This leads to some "leakage" of the incident wave into the scattered field region, which can cause error in the simulation. Since the phase error grows with the propagation distance, this leakage can grow with problem dimension.

A number of methods have been introduced to reduce the leakage by modeling the incident wave with the numerical dispersive wave number of the FDTD grid. Taflove originally proposed to do this with an auxiliary 1-dimensional FDTD grid directed along the axis of propagation to model the incident wave [6]. This was improved somewhat by Guiffaut and Mahdjoubi [7] who proposed to modify the one-dimensional space cell dimension to better match the dispersive properties of the three-dimensional grid. Oguz and Gurel applied signal processing techniques to try to further reduce leakage [8, 9].

The most accurate method to date was proposed by Taflove and Hagness [10] and is referred to as the "Analytical Field Propagation" (AFP) method. An updated version of the AFP algorithm was proposed by Tan and Potter [11]. This method uses Fourier transform theory to compute an exact representation of the analytical field that includes an analytical form for the numerical dispersion of the plane wave. This can be used to provide a highly accurate broad-band representation of the plane wave propagating in the discrete FDTD space.

4.5.5 INHOMOGENEOUS MEDIA

The previous methods assume that the incident wave is propagating in a homogeneous media. It may be desirable to assume that the incident wave is an inhomogeneous media, such as a layered media supporting a printed circuit or antenna. An accurate approach for handling more complex background media was proposed by Watts and Diaz [12]. They refer to their method as "teleportation." In this method, a two or three-dimensional auxiliary grid is used to simulate just the incident field in the complex background media, which is then "teleported" to the actual problem domain with the DUT.

The other alternative is to pose the incident field in the presence of the layered media. The can be solved for analytically *a priori*, and then applied to the total-field/scattered field boundary, in a manner done for the free-space case. The reader is referred to [13]–[16] for details of this approach.

4.6 PROBLEMS

1. You are to excite the lowest order mode (TE10-mode) of a WR-90 rectangular waveguide, which has cross sectional dimensions a = 0.9 inches × 0.4 inches. Find the parameters for a modulated Gaussian pulse, with center frequency fc = 10 GHz such that the amplitude is reduced to 1% of its maximum at the cut-off frequency of the next higher-order TE20 mode.

2. Derive the signature function for a modulated differentiated Gaussian pulse and Blackman-Harris window. Predict the frequency domain response of the signal. Plot out the signal for a 1 GHz center frequency and a 100 MHz bandwidth.

3. A Thévenin voltage source with internal resistance R_{TH} is placed on an x-directed primary grid edge. Derive the first-order update equation for E_x for the source loaded edge.

4. Show that if the plane wave boundary is defined by primary grid cell faces that the updates for the magnetic fields normal to the plane wave boundary do not couple directly to the incident field.

5. Derive the update expressions for the tangential electric fields tangential to the plane wave boundary for all six faces of a rectangular box defining a plane wave boundary. Show that the updates can be expressed as a superposition of the traditional FDTD updates defined by (3.25)–(3.30) with update expressions similar to those defined in (4.47)–(4.50). Explicitly write out these update expressions.

6. Repeat Problem 5, but for the magnetic fields tangential to the plane-wave boundary and one-half cell removed from the boundary.

7. Implement a three-dimensional plane-wave excitation within an existing FDTD computer code. Assume that the plane wave boundary is rectangular in shape, and can be defined by two

vector indices $(i_{min}, j_{min}, k_{min})$ and $(i_{max}, j_{max}, k_{max})$. Set up the code to excite either a vertical or a horizontal polarized plane wave that can propagate along an arbitrary angle of incidence $(\theta^{inc}, \phi^{inc})$. Allow the pulse to have an arbitrary signature defined by the user. To validate the code, set up a problem domain with $\Delta x = \Delta y = \Delta z = 1$ mm. Start with a small space that is $20 \times 20 \times 20$ cells. Define the bounds of the plane-wave boundary to be (3,3,3) and (18,18,18). Apply a Blackmon-Harris window time signature with a 15 GHz BW. Propagate a vertical polarized plane-wave through the FDTD domain with (a) $(\theta^{inc}, \phi^{inc}) = (90°, 0°)$ and (b) $(\theta^{inc}, \phi^{inc}) = (90°, 45°)$. Plot a cross section of the magnitude of the electric field versus time along a cut in the x-z plane. Also, choose a z-directed electric field edge outside of the plane-wave domain to compute the leakage. Choose a position where you expect the leakage to be maximum. Repeat this for a horizontal polarized plane-wave.

REFERENCES

[1] F. J. Harris, "Use of windows for harmonic-analysis with discrete Fourier-transform," *Proceedings of the IEEE*, vol. 66, pp. 51–83, 1978. DOI: 10.1109/PROC.1978.10837 76

[2] Y. H. Chen, W. C. Chew, and M. L. Oristaglio, "Application of perfectly matched layers to the transient modeling of subsurface EM problems," *Geophysics*, vol. 62, pp. 1730–1736, Nov-Dec 1997. DOI: 10.1190/1.1444273 76

[3] V. A. Thomas, M. E. Jones, M. J. Piket-May, A. Taflove, and E. Harrigan, "The use of SPICE lumped circuits as sub-grid models for FDTD high-speed electronic circuit design," *IEEE Microwave and Guided Wave Letters*, vol. 4, pp. 141–143, 1994. DOI: 10.1109/75.289516 89

[4] D. E. Merewether, R. Fisher, and F. W. Smith, "On implementing a numeric Huygen's source scheme in a finite difference program to illuminate scattering bodies," *IEEE Transactions on Nuclear Science*, vol. 27, pp. 1829–1833, 1980. DOI: 10.1109/TNS.1980.4331114 89

[5] G. B. Arfken and J. H. Weber, *Mathematical Methods for Physicists*, 3rd ed. San Diego, CA: Academic Press, 1995. 95

[6] A. Taflove, *Computational Electrodynamics - The Finite Difference Time-Domain Method*. Boston, MA: Artech House, 1995. 97

[7] C. Guiffaut and K. Mahdjoubi, "A perfect wideband plane wave injector for FDTD method " in *IEEE International Symposium on Antennas and Propagation*, Salt Lake City, UT, 2000, pp. 236–239. DOI: 10.1109/APS.2000.873752 97

[8] U. Oguz and L. Gurel, "Reducing the dispersion errors of the finite-difference time-domain method for multifrequency plane-wave excitations," *Electromagnetics*, vol. 23, pp. 539–552, Aug-Sep 2003. DOI: 10.1080/02726340390222062 97

[9] U. Oguz, L. Gurel, and O. Arikan, "An efficient and accurate technique for the incident-wave excitations in the FDTD method," *IEEE Transactions on Microwave Theory and Techniques,* vol. 46, pp. 869–882, Jun 1998. DOI: 10.1109/22.681215 97

[10] A. Taflove and S. B. Hagness, *Computational Electrodynamics: The Finite-Difference Time-Domain,* 3rd ed. Boston, MA: Artech House, 2005. 97

[11] T. Tan and M. Potter, "Optimized analytic field propagator (O-AFP) for plane wave injection in FDTD simulations," *IEEE Transactions on Antennas and Propagation,* vol. 58, pp. 824–831, March 2010. DOI: 10.1109/TAP.2009.2039310 97

[12] M. E. Watts and R. E. Diaz, "Perfect plane-wave injection into a finite FDTD domain through teleportation of fields," *Electromagnetics,* vol. 23, pp. 187–201, Feb-Mar 2003. DOI: 10.1080/02726340390159504 98

[13] I. R. Capoglu, "Techniques for handling multilayered media in the FDTD method," in *School of Electrical and Computer Engineering.* vol. PhD Atlanta: Georgia Institute of Technology, 2007. 98

[14] S. C. Winton, P. Kosmos, and C. Rappaport, "FDTD simulation of TE and TM plane waves at nonzero incidence in arbitrary layered media," *IEEE Transactions on Antennas and Propagation,* vol. 53, pp. 1721–1728, 2005. DOI: 10.1109/TAP.2005.846719

[15] I. R. Capoglu and G. S. Smith, "A total-field/scattered-field plane-wvae source for the FDTD analysis of layered media," *IEEE Transactions on Antennas and Propagation,* vol. 56, pp. 158–169, 2008. DOI: 10.1109/TAP.2007.913088

[16] Y.-N. Jiang, D. B. Ge, and S.-J. Ding, "Analysis of TF-SF boundary for 2D-FDTD with plane P-wave propagation in layered dispersive and lossy media," *Progress in Electromagnetics Research-Pier,* vol. 83, pp. 157–172, 2008. DOI: 10.2528/PIER08042201 98

CHAPTER 5

Absorbing Boundary Conditions

5.1 INTRODUCTION

The challenge that arises when using the FDTD method to simulate electromagnetic fields in un-bounded domains is how to represent an unbounded region with a finite dimensional grid. The space must be truncated with an artificial boundary, which supports a special boundary condition that effectively absorbs any electromagnetic field or wave that impinges upon the boundary. Ideally, all the energy is absorbed, and no energy is reflected back into the problem domain. Such a boundary condition is referred to as an "Absorbing Boundary Condition" (ABC). There are a variety of absorbing boundary conditions (ABCs) that have been developed for this purpose. In general, ABCs can be sorted into three categories: i) local ABCs, ii) global ABCs, and iii) absorbing media. Global ABCs rely on integral-differential operators to compute the fields on the truncation boundary. The tangential fields on (or near) the truncation boundary are treated as equivalent current densities [1]. The fields on the truncation boundary can then be predicted from the equivalent currents using Green's theorems [1, 2]. These operators are global in nature since to compute the field at a single point on the boundary requires an integration over the entire truncation boundary surface. It also requires knowledge of the full time-history of the fields on the boundary. While this is an extremely expensive operation, a number of efficient methods have been introduced to employ such global ABCs [3]–[6]. Nevertheless, at this time, these methods are quite expensive to employ for general purpose problems.

Absorbing media are quite different in that they terminate the problem domain with a physical layer of material absorber, much like an anechoic chamber. In this way, they provide a sponge layer, rather than an explicit boundary condition. The most successful of these methods is the Perfectly Matched Layer (PML) absorbing medium [7], which will be detailed in Chapter 6.

Local absorbing boundary conditions are quite different. These methods are based on approximations of the local derivatives of the fields on the truncation boundary derived from local differential operators that the fields are expected to satisfy. The advantage of this approach is that the operators typically only require knowledge of local fields in the vicinity of the truncation boundary. Therefore, they are extremely efficient to implement. The disadvantage of local ABCs is that they suffer from larger reflection errors as compared to global ABCs or the PML medium.

There are a plethora of ABCs that have been developed for the FDTD method. Some of the more successful methods have been the Engquist-Majda ABC based on the one-wave wave-equation [8]. G. Mur [9] applied the first and second-order Engquist-Majda ABC to the FDTD method. The second-order formulation is now commonly referred to as the Mur-ABC [10].

Trefethen-Halpern added further improvements to this class of ABCs by viewing the ABC as a filter and expressing it as a transfer function [11, 12]. Using a Pade approximation, they were able to generate higher-order models. The limitations of the Engquist-Majda class of ABCs is that they require transverse derivatives on the truncation boundary. As a consequence, these methods require special conditions near the corners of the FDTD lattice grid. These methods are also difficult to apply in more complex media, such as layered media. An approach developed by Higdon [13, 14] helps to over comes these limitations. Betz and Mittra enhanced Higdon's ABC to improve the absorption of evanescent waves and to eliminate late time-instabilities that can occur [15]. Finally, Ramahii combined a succession of absorbing boundaries and suggested running complementary problems in order to cancel errors inherent to the local absorbing boundary conditions, deriving what is referred to as the concurrent complementary operators method (C-COM) [16, 17]. While one could dedicate an entire text book to studying the plethora of absorbing boundary conditions that have been derived, only a sampling is presented in this chapter. Initially, the simple first-order Sommerfeld ABC is introduced. While this is the least accurate, it provides a starting point for understanding the nature of ABC's. The Higdon ABC is chosen as the higher-order ABC to discuss, rather than the Mur ABC. The reason for this is that Higdon's ABC is applicable to homogeneous media as well as layered media. It is also easily extendable to higher-orders. The Betz-Mittra correction to the Higdon ABC is also presented, which allows the ABC to absorb evanescent waves and to provide late time stability.

5.2   THE FIRST-ORDER SOMMERFELD ABC

Consider a domain Ω that is terminated by the ABC boundary $\partial\Omega^{ABC}$, as illustrated in Fig. 5.1. Assume that Ω is discretized using a Yee-based discretization scheme presented in Chapter 3. Assume

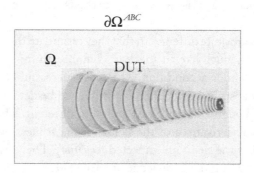

Figure 5.1: Absorbing boundary $\partial\Omega^{ABC}$ terminating the FDTD domain.

that the exterior boundary $\partial\Omega^{ABC}$ is defined by the primary grid (c.f., Fig. 5.1). Consequently, the electric fields tangential to $\partial\Omega^{ABC}$ are sampled on $\partial\Omega^{ABC}$ at the primary grid edge centers, and the magnetic fields normal to $\partial\Omega^{ABC}$ are sampled at the center of the primary grid faces on $\partial\Omega^{ABC}$.

The difficulty that arises is that when updating the tangential electric fields on $\partial\Omega^{ABC}$ using (3.28), (3.29), or (3.30). the expression relies on magnetic fields that are *outside* of $\partial\Omega^{ABC}$, and hence are not known. The magnetic field one-half cell into Ω is known since it is part of the discrete space. However, the magnetic field one-half cell outside of Ω is not known. In other words, one cannot evaluate the normal derivative of the magnetic field tangential to $\partial\Omega^{ABC}$ on the boundary itself. As a consequence, a different operator must be introduced to update the tangential electric field on $\partial\Omega^{ABC}$. This can be done with the use of an absorbing boundary condition operator. In this section, the first-order Sommerfeld ABC is introduced for this purpose.

Consider an outward traveling wave impinging upon the ABC boundary from within Ω that is described as:

$$W(\vec{r}, t) = f(ct - \hat{k} \cdot \vec{r}) \tag{5.1}$$

where c is the speed of light and $\hat{k}$ is the direction of propagation. For simplicity, assume the wave is propagating in the x-direction and is normally incident upon the boundary, that is, $\hat{k} = \hat{n} = \hat{x}$. Consequently, (5.1) can be written as:

$$W(x, t) = f(ct - x) \tag{5.2}$$

where n is the coordinate along the normal axis. If the function f is at least once differentiable, then it can be shown that $W(x, t)$ will satisfy the differential operator:

$$\left(\frac{\partial}{\partial t} + c\frac{\partial}{\partial x}\right) W(x, t) = 0. \tag{5.3}$$

Now, consider a wave that is arbitrarily incident upon the boundary, with the direction of propagation $\hat{k}$ making an angle ϕ off of the normal axis as illustrated in Fig. 5.2. For simplicity, assume that $\hat{n} = \hat{x}$

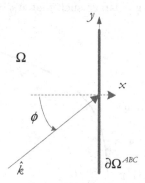

Figure 5.2: Plane wave impinging on $\partial\Omega^{ABC}$ with angle ϕ off the normal.

and that the boundary lies in the $x = 0$ plane. Using (5.3), one can approximate:

$$\left(\frac{\partial}{\partial t} + c\frac{\partial}{\partial x}\right) W(\hat{r}, t) \approx 0. \tag{5.4}$$

The approximation is exact if $\hat{k} = \hat{x}$. Otherwise, there is error in the approximation. The error is manifested as a spurious reflection. For sake of illustration, assume that the wave is propagating in the x, y-plane. Then,

$$W^{inc}(\vec{r}, t) = f(ct - \hat{k} \cdot \vec{r}) = f(ct - x \cos\phi - y \sin\phi). \tag{5.5}$$

The reflected wave can then be posed as:

$$W^{ref}(\vec{r}, t) = Rf(ct + x \cos\phi - y \sin\phi) \tag{5.6}$$

where R is the reflection coefficient. Note that the sign of x is flipped since the reflected wave is traveling away from the boundary. The total field is $W = W^{inc} + W^{ref}$. From (5.4), we can express:

$$\left(\frac{\partial}{\partial t} + c\frac{\partial}{\partial x}\right)\left(W^{inc}(\vec{r}, t) + W^{ref}(\vec{r}, t)\right)\bigg|_{x=0} = 0. \tag{5.7}$$

Expanding out the derivatives, this leads to:

$$(c - c \cos\phi + R(c + c\cos\phi)) f'(ct - y \sin\phi) = 0 \tag{5.8}$$

where f' is the derivative of f with respect to its argument. Solving for the reflection coefficient R:

$$R = \frac{\cos\phi - 1}{\cos\phi + 1}. \tag{5.9}$$

The magnitude of the reflection error is illustrated in Fig. 5.3. At normal incidence ($\phi = 0$), the reflection error is 0. However, as the angle of incidence deviates further from normal incidence, the reflection error becomes increasingly larger, such that at $\phi = 90°$, $R = 1$.

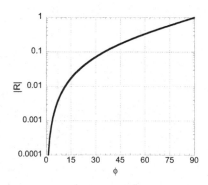

Figure 5.3: Reflection error of the first-order Sommerfeld ABC.

The Sommerfeld ABC can be used to compute the tangential electric fields in the plane of the ABC. Again, let the local outward normal of $\partial\Omega^{ABC}$ be along the x-direction. Then, consider the primary edge E_y on $\partial\Omega^{ABC}$. The ABC is expressed as:

$$\left(\frac{\partial}{\partial t} + c\frac{\partial}{\partial x}\right) E_y\left(\hat{r}, t\right)\bigg|_{\partial\Omega^{ABC}} = 0. \tag{5.10}$$

The x-derivative of E_y is computed a half of a space cell inside the boundary. If the local ABC boundary is defined by the $i = N_x$ boundary in the discrete space, then (5.10) is approximated via central difference approximations as

$$\frac{1}{2}\left(\frac{E_y^{n+\frac{1}{2}}\Big|_{N_x, j+\frac{1}{2}, k} - E_y^{n-\frac{1}{2}}\Big|_{N_x, j+\frac{1}{2}, k}}{\Delta t} + \frac{E_y^{n+\frac{1}{2}}\Big|_{N_x-1, j+\frac{1}{2}, k} - E_y^{n-\frac{1}{2}}\Big|_{N_x-1, j+\frac{1}{2}, k}}{\Delta t}\right) +$$

$$\frac{c}{2}\left(\frac{E_y^{n+\frac{1}{2}}\Big|_{N_x, j+\frac{1}{2}, k} - E_y^{n+\frac{1}{2}}\Big|_{N_x-1, j+\frac{1}{2}, k}}{\Delta x} + \frac{E_y^{n-\frac{1}{2}}\Big|_{N_x, j+\frac{1}{2}, k} - E_y^{n-\frac{1}{2}}\Big|_{N_x-1, j+\frac{1}{2}, k}}{\Delta x}\right) = 0. \tag{5.11}$$

To maintain second order accuracy at position $(N_x - \frac{1}{2})\Delta x$ and at time $n\Delta t$, the first term in (5.11) has been averaged in space, and the second term averaged in time. This expression leads to the a recursive update expression:

$$E_y^{n+\frac{1}{2}}\Big|_{N_x, j+\frac{1}{2}, k} = \gamma E_y^{n-\frac{1}{2}}\Big|_{N_x, j+\frac{1}{2}, k} - \gamma E_y^{n+\frac{1}{2}}\Big|_{N_x-1, j+\frac{1}{2}, k} + E_y^{n-\frac{1}{2}}\Big|_{N_x-1, j+\frac{1}{2}, k} \tag{5.12}$$

where

$$\gamma = \frac{1-\rho}{1+\rho}, \text{ and } \rho = \frac{c\Delta t}{\Delta x}. \tag{5.13}$$

The recursive relationship (5.12) can be used to update the tangential electric fields on the exterior boundary. This expression is stable within the Courant stability limit of the FDTD method. The update relies on the interior field at time-step $n + \frac{1}{2}$. Thus, the interior fields must be updated first, using the standard recursive relationship (3.29), prior to updating the boundary fields. Also, note that $E_y^{n-\frac{1}{2}}\Big|_{N_x-1, j+\frac{1}{2}, k}$ has to be back-stored. Finally, once the electric fields are updated, the magnetic fields normal to $\partial\Omega^{ABC}$ can be updated normally using (3.25)–(3.27).

5.3   THE HIGDON ABC

The Sommerfeld boundary condition provides a first-order ABC. For most applications, the reflection error (c.f., Fig. 5.3) is excessively high. This can be improved by applying higher-order ABC's.

A viable alternative is the Higdon boundary operator [13, 14]. To better understand the Higdon ABC, consider the field incident on the absorbing boundary to be a spectrum of plane waves:

$$W(\vec{r}, t) = \sum_i f_i \left(ct - \hat{k}_i \cdot \vec{r} \right) \tag{5.14}$$

where, for simplicity, we assume that:

$$\hat{k}_i = \cos \phi_i \hat{x} + \cos \phi_i \hat{y}. \tag{5.15}$$

Higdon proposed the following *annihilation* operator:

$$\mathcal{H}\left(\phi_j; j = 1..N \right) = \prod_{j=1}^{N} \left(\frac{\partial}{\partial t} + \frac{c}{\cos \phi_j} \frac{\partial}{\partial x} \right) \tag{5.16}$$

such that

$$\mathcal{H}W(\vec{r}, t) = 0 \tag{5.17}$$

if the $\phi_j = \phi_i$. In general, the ϕ_i are not known. Consequently, one chooses ϕ_j in order to cover a spectrum of plane waves.

The order of the Higdon boundary operator is determined by the number of angles N. For example, if $N = 1$, and $\phi_1 = 0$, the Higdon boundary operator is identical to the Sommerfeld boundary condition. For the second-order case, $N = 2$, the Higdon boundary operator becomes:

$$\left(\frac{\partial}{\partial t} + \frac{c}{\cos \phi_1} \frac{\partial}{\partial x} \right) \left(\frac{\partial}{\partial t} + \frac{c}{\cos \phi_2} \frac{\partial}{\partial x} \right) W(\vec{r}, t) = 0. \tag{5.18}$$

The reflection error of a single plane wave can be determined by expanding $W(\vec{r}, t)$ as a superposition of incident and reflected waves, as in (5.5) and (5.6). Applying the second-order Higdon boundary operator (5.18), results in the reflection coefficient

$$R(\phi) = -\frac{\cos \phi_1 \cos \phi_2 - \cos \phi_1 \cos \phi - \cos \phi_2 \cos \phi + \cos^2 \phi}{\cos \phi_1 \cos \phi_2 + \cos \phi_1 \cos \phi + \cos \phi_2 \cos \phi + \cos^2 \phi}. \tag{5.19}$$

The magnitude of the reflection error is illustrated in Fig. 5.4 for a variety of angle pairs (ϕ_1, ϕ_2). Note that when $(\phi_1, \phi_2) = (0°, 0°)$, the reflection error is identical to that of a second-order Mur ABC [9]. For this pair, the normal incident wave is annihilated with a second-order operator. This broadens the depth of the null as compared to the first-order ABC illustrated in Fig. 5.3. By choosing a different pair, such as $(\phi_1, \phi_2) = (10°, 27.5°)$, the bandwidth can be increased, as seen from Fig. 5.4. The tradeoff is that the reflection error increases near normal incidence. However, the overall reflection error can be maintained to be < 0.001 for all $\phi < 30°$. For the angle pair $(\phi_1, \phi_2) = (15°, 45°)$, the reflection error increases again. However, the overall reflection error can be maintained to be < 0.01 for all $\phi < 50°$. Thus, one can synthesize the response of the ABC

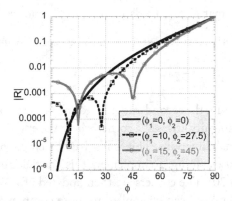

Figure 5.4: Reflection error for the second-order Higdon ABC.

by the choice of the angle pair (ϕ_1, ϕ_2). With the Higdon operator, one can increase the angular spectrum while sacrificing the level of reflection error.

In the discrete space, the second-order Higdon ABC is used to update the tangential fields on the ABC boundary. Again, as an example, assume that the absorbing boundary is defined via the $i = N_x$ plane. The difference operator is derived from (5.18) by using a product rule for the derivatives. Initially, the difference operator for $(\frac{\partial}{\partial t} + \frac{c}{\cos \phi_2} \frac{\partial}{\partial x})$ is evaluated at position $(N_x - \frac{1}{2})\Delta x$ and time $n\Delta t$, as done in (5.11). A similar expression is derived at position $(N_x - \frac{3}{2})\Delta x$. Similar expressions are derived to be second-order accurate at time $(n-1)\Delta t$. Next, the difference operator $(\frac{\partial}{\partial t} + \frac{c}{\cos \phi_1} \frac{\partial}{\partial x})$ is evaluated by operating on these previously derived discrete equations. This is done in a manner that is second-order accurate at $(N_x - 1)\Delta x$ and time $(n - \frac{1}{2})\Delta t$ using central differencing and averaging. From this, the discrete update for E_y on the ABC boundary is defined as:

$$
\begin{aligned}
E_y^{n+\frac{1}{2}}\Big|_{N_x, j+\frac{1}{2}, k} = {} & -(\gamma_1 + \gamma_2) E_y^{n+\frac{1}{2}}\Big|_{N_x-1, j+\frac{1}{2}, k} - \gamma_1\gamma_2 E_y^{n+\frac{1}{2}}\Big|_{N_x-2, j+\frac{1}{2}, k} + (\gamma_1 + \gamma_2) E_y^{n-\frac{1}{2}}\Big|_{N_x, j+\frac{1}{2}, k} \\
& + (2 + 2\gamma_1\gamma_2) E_y^{n-\frac{1}{2}}\Big|_{N_x-1, j+\frac{1}{2}, k} + (\gamma_1 + \gamma_2) E_y^{n-\frac{1}{2}}\Big|_{N_x-2, j+\frac{1}{2}, k} \\
& - \gamma_1\gamma_2 E_y^{n-\frac{3}{2}}\Big|_{N_x, j+\frac{1}{2}, k} - (\gamma_1 + \gamma_2) E_y^{n-\frac{3}{2}}\Big|_{N_x-1, j+\frac{1}{2}, k} - E_y^{n-\frac{3}{2}}\Big|_{N_x-2, j+\frac{1}{2}, k}
\end{aligned}
\tag{5.20}
$$

where,

$$
\gamma_i = \frac{1 - \rho_i}{1 + \rho_i}, \quad \rho_i = \frac{c\Delta t}{\cos \phi_i \Delta x}.
\tag{5.21}
$$

This local recursive operator is used to update the tangential electric field on the ABC boundary. This operator is stable within the Courant stability limit of the FDTD algorithm. The proper implementation assumes that all the fields interior to Ω, and excluding $\partial\Omega^{ABC}$ are first updated using (3.29). It is also noted that special storage must be used to back store $E_y^{n-\frac{1}{2}}\Big|_{N_x-1,j+\frac{1}{2},k}$, $E_y^{n-\frac{1}{2}}\Big|_{N_x-2,j+\frac{1}{2},k}$, $E_y^{n-\frac{3}{2}}\Big|_{N_x,j+\frac{1}{2},k}$, $E_y^{n-\frac{3}{2}}\Big|_{N_x-1,j+\frac{1}{2},k}$, and $E_y^{n-\frac{3}{2}}\Big|_{N_x-2,j+\frac{1}{2},k}$.

The stencil of points for the second-order Higdon boundary operator lie on a single axis normal to the ABC plane. An ambiguity arises at a corner where the edge is shared by two intersecting ABC planes with distinct normal directions. This is resolved by employing a Higdon boundary operator relative to both planes and averaging. To implement this, the fields on the plane excluding the corners are first updated using (5.20). The corners are then updated afterwards in a separate loop.

Finally, the Higdon ABC can be applied to more complex media, such as layered media. In practice, the choice for c is determined by the local material constants.

5.4   THE BETZ-MITTRA ABC

When an electromagnetic wave is interacting with an arbitrary geometry, an entire spectrum of waves will be scattered, including both propagating and *evanescent* waves. The Higdon ABC is designed to annihilate propagating waves, but is deficient at absorbing evanescent waves. As a consequence, the ABC must be pushed sufficiently far away from the device under test such that the evanescent waves have adequately decayed prior to interacting with the ABC. Since it must be extended in multiple directions, this leads to a significant increase in the size of the global grid. To circumvent this problem, Betz and Mittra proposed a form of the Higdon ABC that also annihilates evanescent waves [15]. For a second-order ABC, Betz and Mittra proposed:

$$\left(\frac{\partial}{\partial t} + \frac{c}{\cos\phi_1}\frac{\partial}{\partial x} + \alpha_1\right)\left(\frac{\partial}{\partial t} + \frac{c}{\cos\phi_2}\frac{\partial}{\partial x}\right)W(\vec{r},t) = 0 \qquad (5.22)$$

where α_1 is an attenuation constant, and the boundary is assumed to have an x-directed normal. This term is designed to annihilate waves of the form $W(\vec{r},t) = e^{-\vec{\gamma}\cdot\vec{r}}e^{j\omega t}$, where $\gamma_x = jk\cos\phi_i + \alpha_i$, and ω is a radial frequency.

The second-order Betz-Mittra ABC is discretized in a similar manner as the second-order Higdon ABC, leading to a recursive operator that can be used to time-advance the fields on the ABC boundary. For example, the recursive update operator for E_y on the $i = N_x$ boundary is:

$$\begin{aligned}
E_y^{n+\frac{1}{2}}\Big|_{N_x,j+\frac{1}{2},k} = &-(\gamma_1+\gamma_2)E_y^{n+\frac{1}{2}}\Big|_{N_x-1,j+\frac{1}{2},k} - \gamma_1\gamma_2 E_y^{n+\frac{1}{2}}\Big|_{N_x-2,j+\frac{1}{2},k} + (\gamma_1+\gamma_2)E_y^{n-\frac{1}{2}}\Big|_{N_x,j+\frac{1}{2},k} \\
&+ ((\beta+1)+2\gamma_1\gamma_2)E_y^{n-\frac{1}{2}}\Big|_{N_x-1,j+\frac{1}{2},k} + (\beta\gamma_2+\gamma_1)E_y^{n-\frac{1}{2}}\Big|_{N_x-2,j+\frac{1}{2},k} \\
&- \gamma_1\gamma_2 E_y^{n-\frac{3}{2}}\Big|_{N_x,j+\frac{1}{2},k} - (\beta\gamma_2+\gamma_1)E_y^{n-\frac{3}{2}}\Big|_{N_x-1,j+\frac{1}{2},k} - \beta E_y^{n-\frac{3}{2}}\Big|_{N_x-2,j+\frac{1}{2},k}
\end{aligned} \qquad (5.23)$$

where

$$\gamma_i = \frac{1 - \rho_i}{1 + \rho_i(1 + \alpha_1 \Delta x)}, \quad \beta = \frac{1 + \rho_i}{1 + \rho_i(1 + \alpha_1 \Delta x)}, \quad \rho_i = \frac{c\Delta t}{\cos \phi_i \Delta x}. \quad (5.24)$$

The annihilation angles (ϕ_1, ϕ_2) are chosen in a similar manner as the Higdon ABC. A typical value for the attenuation constant is chosen to be $\alpha_1 \Delta x = 0.1$ [15].

The recursive update scheme in (5.23) provides a second-order accurate representation of (5.22). The framework of the update is identical to the second-order Higdon ABC. The only difference is the coefficients. The advantage of this operator is that it can be moved closer to the DUT since it also absorbs some of the evanescent wave energy. Another advantage of this scheme is that in some instances the second-order Higdon ABC can suffer from late-time instability. This is circumvented with the application of the Betz-Mittra ABC.

5.5   PROBLEMS

1. Show that if $W(x, t)$ is defined by (5.2) and $W(x, t)$ is differentiable, then (5.3) is exactly true. (Hint: Differentiate f with respect to its argument and apply the chain rule to compute the derivatives with respect to t and x.)

2. Derive the reflection error for a third-order Higdon boundary operator. Plot the reflection error versus ϕ when (ϕ_1, ϕ_2, ϕ_3) = ($0°, 0°, 0°$) and ($10°, 27.5°, 45°$).

3. Derive (5.20).

4. Derive expressions similar to (5.20) for the ABC boundary for the discrete plane $i = 1$.

5. Derive expressions similar to (5.20) for updating E_x and E_z.

6. Derive the update expression in (5.23).

REFERENCES

[1] A. F. Peterson, S. L. Ray, and R. Mittra, *Computational Methods for Electromagnetics*. New York: IEEE Press, 1998. 101

[2] C. A. Balanis, *Advanced Engineering Electromagnetics*. New York: Wiley, 1989. 101

[3] R. Holtzman and R. Kastner, "The time-domain discrete Green's function method (GFM) characterizing the FDTD grid boundary," *IEEE Transactions on Antennas and Propagation*, vol. 49, pp. 1079–1093, Jul 2001. DOI: 10.1109/8.933488 101

[4] W. L. Ma, M. R. Rayner, and C. G. Parini, "Discrete Green's function formulation of the FDTD method and its application in antenna modeling," *IEEE Transactions on Antennas and Propagation*, vol. 53, pp. 339–346, Jan 2005. DOI: 10.1109/TAP.2004.838797

[5] N. Rospsha and R. Kastner, "Closed form FDTD-compatible Green's function based on combinatorics," *Journal of Computational Physics,* vol. 226, pp. 798–817, Sep 2007. DOI: 10.1016/j.jcp.2007.05.017

[6] J. Vazquez and C. G. Parini, "Discrete Green's function formulation of FDTD method for electromagnetic modelling," *Electronics Letters,* vol. 35, pp. 554–555, Apr 1999. DOI: 10.1049/el:19990416 101

[7] J.-P. Berenger, "A perfectly matched layer for the absorption of electromagnetic waves," *Journal of Computational Physics,* vol. 114, pp. 195–200, 1994. DOI: 10.1006/jcph.1994.1159 101

[8] B. Engquist and A. Majda, "Absorbing boundary-conditions for numerical simulation of waves," *Mathematics of Computation,* vol. 31, pp. 629–651, 1977. DOI: 10.1090/S0025-5718-1977-0436612-4 101

[9] G. Mur, "Absorbing boundary-conditions for the finite-difference approximation of the time-domain electromagnetic-field equations," *IEEE Transactions on Electromagnetic Compatibility,* vol. 23, pp. 377–382, 1981. DOI: 10.1109/TEMC.1981.303970 101, 106

[10] A. Taflove and S. B. Hagness, *Computational Electrodynamics: The Finite-Difference Time-Domain,* 3rd ed. Boston, MA: Artech House, 2005. 101

[11] L. Halpern and L. N. Trefethen, "Wide-angle one-wave wave-equations," *Journal of the Acoustical Society of America,* vol. 84, pp. 1397–1404, Oct 1988. DOI: 10.1121/1.396586 102

[12] L. N. Trefethen and L. Halpern, "Well-posedness of one-wave wave-equations and absorbing boundary-conditions," *Mathematics of Computation,* vol. 47, pp. 421–435, Oct 1986. DOI: 10.1090/S0025-5718-1986-0856695-2 102

[13] R. L. Higdon, "Absorbing boundary-conditions for difference approximations to the multidimensional wave-equation " *Mathematics of Computation,* vol. 47, pp. 437–459, Oct 1986. DOI: 10.1090/S0025-5718-1986-0856696-4 102, 106

[14] R. L. Higdon, "Numerical absorbing boundary-conditions for the wave-equation," *Mathematics of Computation,* vol. 49, pp. 65–90, Jul 1987. DOI: 10.1090/S0025-5718-1987-0890254-1 102, 106

[15] V. Betz and R. Mittra, "Comparison and evaluation of boundary conditions for the absorption of guided waves in an FDTD simulation," *IEEE Microwave and Guided Wave Letters,* vol. 2, pp. 499–501, 1992. DOI: 10.1109/75.173408 102, 108, 109

[16] O. M. Ramahi, "The complementary operators method in FDTD simulations," *IEEE Antennas and Propagation Magazine,* vol. 39, pp. 33–45, Dec 1997. DOI: 10.1109/74.646801 102

[17] O. M. Ramahi, "The concurrent complementary operators method for FDTD mesh truncation," *IEEE Transactions on Antennas and Propagation*, vol. 46, pp. 1475–1482, Oct 1998. DOI: 10.1109/8.725279 102

CHAPTER 6

The Perfectly Matched Layer (PML) Absorbing Medium

6.1   INTRODUCTION

For three decades, unbounded problem domains simulated via the FDTD method were routinely terminated by local absorbing boundary conditions (ABCs). Due to the relatively high reflection errors encountered with local ABCs, the truncation boundary had to be extended sufficiently far off such that evanescent near-fields were amply decayed, and so that waves would not impinge on the boundary near grazing incidence. These ABCs were also quite inept at terminating complex media, such as anisotropic, lossy, dispersive, or inhomogeneous media. This all changed in the mid 1990's with the advent of the *perfectly matched layer* absorbing media, pioneered by Jeanne-Pierre Bérenger [2, 3]. The perfectly matched layer, better known as PML, is so-called since an arbitrarily polarized, broad-band electromagnetic wave impinging on a PML half-space will penetrate into the medium without reflection. The constitutive parameters of the PML are selected such that waves entrant into the medium rapidly decay. With such properties, the PML provides a highly effective means to terminate a FDTD lattice in the same manner as one would terminate an anechoic chamber with absorbing materials. Unlike local ABCs, the PML can absorb both propagating and evanescent waves. As a consequence, the outer truncation boundary can be placed much closer to the device under test. The PML can also be perfectly matched to complex media, such as lossy, dispersive, inhomogeneous media, and even non-linear media. Needless to say, the PML absorbing media boundary has enabled the FDTD to become an effective simulation tool for a wide range of applications.

Bérenger originally contrived the PML using a novel split-field formulation of Maxwell's equations [2]. Soon after, it was shown that Bérenger's PML was identical to expressing Maxwell's equations in a complex-stretched coordinate frame [4]. It was found that the PML could also be posed as an anisotropic medium [5, 6].

In this chapter, an introduction to the properties of a perfectly matched layer absorbing medium is provided. The application of the PML as an absorbing layer for the FDTD method is also presented. Rather than providing an exhausted survey of all the manifestations of the PML [3], Roden's Convolutional-PML (C-PML) method [7] is applied to the FDTD method with the Complex-Frequency-Shifted (CFS) PML constitutive parameters. To date, this is the most robust and computationally efficient method for terminating a FDTD grid.

6.2   THE ANISOTROPIC PML

Consider an arbitrarily polarized time-harmonic plane wave propagating in an isotropic half-space Ω_1 ($x < 0$) (c.f., Fig. 6.1), assumed to be of the form:

$$\vec{H}^{\text{inc}} = \vec{H}_0 \, e^{-jk_{1_x}x - jk_{1_y}y} \tag{6.1}$$

with propagation vector

$$\vec{k} = \hat{x}k_{1_x} + \hat{y}k_{1_y} = \hat{x}k_1 \cos\phi^{inc} + \hat{y}k_1 \sin\phi^{inc}, \tag{6.2}$$

$$k_1 = \omega\sqrt{\varepsilon_1\mu_1}, \tag{6.3}$$

where a $e^{j\omega t}$ time-dependence has been assumed. This wave is assumed to be incident on the half-

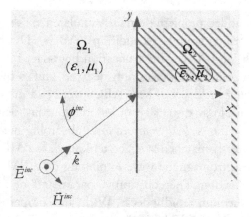

Figure 6.1: Uniform plane wave incident on the uniaxial medial half space.

space Ω_2 ($x > 0$), that is a uniaxial anisotropic medium having the permittivity and permeability tensors:

$$\bar{\bar{\varepsilon}}_2 = \varepsilon_2 \begin{bmatrix} a & 0 & 0 \\ 0 & b & 0 \\ 0 & 0 & b \end{bmatrix}, \qquad \bar{\bar{\mu}}_2 = \mu_2 \begin{bmatrix} c & 0 & 0 \\ 0 & d & 0 \\ 0 & 0 & d \end{bmatrix}. \tag{6.4}$$

The anisotropy is assumed to be rotationally symmetric about the x-axis (that is, $\varepsilon_{yy} = \varepsilon_{zz} = \varepsilon_2 b$ and $\mu_{yy} = \mu_{zz} = \mu_2 d$). The wave transmitted into Ω_2 must satisfy Maxwell's equations, which can be expressed as:

$$-j\omega\bar{\bar{\mu}}_2\vec{H} = \nabla \times \vec{E}, \qquad j\omega\bar{\bar{\varepsilon}}_2\vec{E} = \nabla \times \vec{H}. \tag{6.5}$$

The wave transmitted into Ω_2 is also a plane wave, and the fields will have the dependence $e^{-j\vec{k}_2 \cdot \vec{r}}$, where $\vec{k}_2 = \hat{x}k_{2_x} + \hat{y}k_{2_y}$ is the propagation vector in Ω_2 which still must be determined. Consequently, (6.5), can be written as:

$$\omega\bar{\bar{\mu}}_2\vec{H} = \vec{k}_2 \times \vec{E}, \qquad \omega\bar{\bar{\varepsilon}}_2\vec{E} = -\vec{k}_2 \times \vec{H}. \tag{6.6}$$

Applying the operator $\vec{k}_2 \times \bar{\bar{\varepsilon}}_2^{-1}$ to the second equation of (6.6), and then substituting for $\vec{k}_2 \times \vec{E}$ from the first equation leads to the wave equation:

$$\vec{k}_2 \times \bar{\bar{\varepsilon}}_2^{-1} \vec{k}_2 \times \vec{H} + \omega^2 \bar{\bar{\mu}}_2 \vec{H} = 0. \tag{6.7}$$

Expanding out the cross product, and substituting in (6.4), (6.7) can be expressed in a matrix form as:

$$\begin{bmatrix} k_2^2 c - \left(k_{2_y}\right)^2 b^{-1} & k_{2_x} k_{2_y} b^{-1} & 0 \\ k_{2_x} k_{2_y} b^{-1} & k_2^2 d - \left(k_{2_x}\right)^2 b^{-1} & 0 \\ 0 & 0 & k_2^2 d - \left(k_{2_x}\right)^2 b^{-1} - \left(k_{2_y}\right)^2 a^{-1} \end{bmatrix} \begin{bmatrix} H_x \\ H_y \\ H_z \end{bmatrix} = 0 \tag{6.8}$$

where $k_2 = \omega\sqrt{\varepsilon_2 \mu_2}$. This expression must be true for non-zero magnetic fields. A necessary condition for this is that the determinant of the matrix must be zero. This leads to what is referred to as the dispersion relation. It is observed from the matrix operator that there is a natural decoupling into two distinct polarizations. The first is referred to as the transverse-magnetic to z (TM$_z$) polarization ($H_z = 0$), which is governed by the 2×2 matrix block in the upper left. The second is the transverse-electric to z (TE$_z$) polarization ($H_x, H_y = 0$), which is governed by the 1×1 matrix block in the lower left. The determinants of each of these matrix blocks lead to two distinct dispersion relations:

$$k_2^2 - k_{2_x}^2 b^{-1} d^{-1} - k_{2_y}^2 b^{-1} c^{-1} = 0, \qquad \text{TM}_z \ (H_z = 0) \tag{6.9}$$

$$k_2^2 - k_{2_x}^2 b^{-1} d^{-1} - k_{2_y}^2 a^{-1} d^{-1} = 0, \qquad \text{TE}_z \ (H_x, H_y = 0) \tag{6.10}$$

which govern each polarization.

Consider first the case when the incident field is a TE$_z$-polarized. The incident plane wave will reflect off the interface between the two media. The total fields in Ω_1 are then expressed as a superposition of the incident and reflected waves as [8]:

$$\vec{H}_1 = \hat{z} H_0 \left(1 + \Gamma e^{2jk_{1_x} x}\right) e^{-jk_{1_x} x - jk_{1_y} y}$$

$$\vec{E}_1 = \left[-\hat{x} \frac{k_{1_y}}{\omega \varepsilon_1} \left(1 + \Gamma e^{2jk_{1_x} x}\right) + \hat{y} \frac{k_{1_x}}{\omega \varepsilon_1} \left(1 - \Gamma e^{2jk_{1_x} x}\right)\right] H_0 e^{-jk_{1_x} x - jk_{1_y} y} \tag{6.11}$$

where Γ is the reflection coefficient. The wave transmitted into Region 2 is also TE$_z$ with propagation characteristics governed by (6.5). These fields are expressed as

$$\vec{H}_2 = \hat{z} H_0 \tau e^{-jk_{2_x} x - jk_{2_y} y}$$

$$\vec{E}_2 = \left(-\hat{x} \frac{k_{2_y}}{\omega \varepsilon_2 a} + \hat{y} \frac{k_{2_x}}{\omega \varepsilon_2 b}\right) H_0 \tau e^{-jk_{2_x} x - jk_{2_y} y} \tag{6.12}$$

and τ is the transmission coefficient. Γ and τ are found by enforcing continuity of the tangential E and H fields across $x = 0$ (c.f., 3.11). Initially, for continuity of all values of y, there must be a phase matching. That is,

$$k_{2_y} = k_{1_y}. \tag{6.13}$$

Enforcing 3.11 at $x = 0$, Γ and τ are found to be:

$$\Gamma = \frac{\dfrac{k_{1_x}}{\varepsilon_1} - \dfrac{k_{2_x}}{\varepsilon_2 b}}{\dfrac{k_{1_x}}{\varepsilon_1} + \dfrac{k_{2_x}}{\varepsilon_2 b}}, \quad \tau = 1 + \Gamma = \frac{2\dfrac{k_{1_x}}{\varepsilon_1}}{\dfrac{k_{1_x}}{\varepsilon_1} + \dfrac{k_{2_x}}{\varepsilon_2 b}}. \tag{6.14}$$

Combining the phase match condition (6.13) with the dispersion relation (6.10):

$$k_{2_x} = \sqrt{k_2^2 b\, d - \left(k_{1_y}\right)^2 a^{-1} b}. \tag{6.15}$$

It is observed that if one chooses $\varepsilon_2 = \varepsilon_1$ and $\mu_2 = \mu_1$, then $k_2 = k_1$. Furthermore, if $d = b$ and $a^{-1} = b$, then

$$k_{2_x} = \sqrt{k_1^2 b^2 - \left(k_{1_y}\right)^2 b^2} = b\sqrt{k_1^2 - \left(k_{1_y}\right)^2} \equiv b\, k_{1_x}. \tag{6.16}$$

If this is substituted into (6.14), then $\Gamma = 0$ for all k_{1_x}! As a result, the interface between the two regions is reflectionless for the TE_z-polarized wave.

This can be repeated for the TM_z-polarized wave. It is found that the TM_z-polarized case also leads to a reflectionless interface if $\varepsilon_2 = \varepsilon_1$, $\mu_2 = \mu_1$, $b = d$ and $c^{-1} = d$.

In summary, the uniaxial medium with material constants (6.4) is perfectly matched to the half space Ω_1 if $\varepsilon_2 = \varepsilon_1$, $\mu_2 = \mu_1$, and $d = a^{-1} = c^{-1} = b$. That is, the incident wave will completely enter into Ω_2 without reflection, and the transmitted wave will propagate with a wave vector $\vec{k}_2 = \hat{x} b k_{1_x} + \hat{y} k_{1_y}$. This result is independent of the angle of incidence, the frequency, the wave polarization, and ε_1 and μ_1.

For the case when the PML media boundary has an x-directed normal, a more convenient notation is to let $b = s_x$. Consequently, from (6.4), the uniaxial media with an x-normal is perfectly matched to the half space (μ_1, ε_1) when:

$$\bar{\bar{\varepsilon}}_2 = \varepsilon_1 \bar{\bar{s}}, \qquad \bar{\bar{\mu}}_2 = \mu_1 \bar{\bar{s}}, \tag{6.17}$$

where

$$\bar{\bar{s}} = \bar{\bar{s}}_x = \begin{bmatrix} s_x^{-1} & 0 & 0 \\ 0 & s_x & 0 \\ 0 & 0 & s_x \end{bmatrix}. \tag{6.18}$$

Based on (6.16), it is found that the wave transmitted into the PML region has a dependence along the normal direction of:

$$e^{-jk_{1_x}s_x x}. \tag{6.19}$$

The PML constitutive parameter s_x scales k_{1_x} and thus its choice will strongly impact the propagation properties of the transmitted wave along the normal direction. Bérenger suggested [2, 3]

$$s_x = \kappa_x + \frac{\sigma_x}{j\omega\varepsilon_o}. \tag{6.20}$$

With this choice:

$$e^{-jk_1 s_x x} = e^{-jk_{1_x}\kappa_x x} e^{-\sqrt{\varepsilon_{r_1}\mu_{r_1}}\eta_o \sigma_x \cos\phi^{inc} x} \tag{6.21}$$

where $k_{1_x} = \omega\sqrt{\varepsilon_1\mu_1}\cos\phi^{inc}$ has been assumed, ε_{r_1} and μ_{r_1} are the relative permittivity and permeability, respectively, and $\eta_o = \sqrt{\mu_o/\varepsilon_o}$ is the free-space wave impedance. For causality, $\sigma_x \geq 1$. Also, one would typically choose $\kappa_x \geq 1$. The imaginary term of s_x leads to an attenuation of the signal. Most importantly, the rate of attenuation is independent of frequency. Therefore, the attenuation is merely a function of the physical distance traveled through the media, and not a function of the wavelength. Also, the real part κ_x essentially inversely scales the effective phase velocity of the wave traveling through the medium.

Thus far, the PML medium being investigated is a half-space separated by a boundary with a normal directed along the x-direction. This same analogy can be applied to half-spaces defined by normals directed along either the y or the z-directions by simply rotating the coordinate axis. To this end, uniaxial media would be defined as (6.17), with

$$\bar{\bar{s}} = \bar{\bar{s}}_y = \begin{bmatrix} s_y & 0 & 0 \\ 0 & s_y^{-1} & 0 \\ 0 & 0 & s_y \end{bmatrix}, \text{ or } \bar{\bar{s}} = \bar{\bar{s}}_z = \begin{bmatrix} s_z & 0 & 0 \\ 0 & s_z & 0 \\ 0 & 0 & s_z^{-1} \end{bmatrix}, \tag{6.22}$$

for y-directed normals and z-directed normals, respectively.

The PML is also perfectly matched to an anisotropic medium. For example, consider Fig. 6.2, where the PML half space, Ω_2, shares a planar boundary with an anisotropic half space Ω_1 having material profile $(\bar{\bar{\varepsilon}}_1, \bar{\bar{\mu}}_1)$. In this example, the boundary has a y-directed normal. Following the same arguments, the PML is perfectly matched to the half space providing that

$$\bar{\bar{\varepsilon}}_2 = \bar{\bar{s}}_y \cdot \bar{\bar{\varepsilon}}_1, \qquad \bar{\bar{\mu}}_2 = \bar{\bar{s}}_y \cdot \bar{\bar{\mu}}_1. \tag{6.23}$$

This is very important because when applying a PML as an absorbing layer to terminate a FDTD grid, the distinct PML layers will be overlapping in the corner regions of the grid. In these regions, it is important to ensure that the overlapping PML layers are perfectly matched. For example, consider Ω_1 in Fig. 6.2 to be a PML region defined with an x-directed normal, and, $\bar{\bar{\varepsilon}}_1 = \bar{\bar{s}}_x \varepsilon_1$ and $\bar{\bar{\mu}}_1 = \bar{\bar{s}}_x \mu_1$. Consequently, from (6.23) if $\bar{\bar{\varepsilon}}_2 = \bar{\bar{s}}_x \cdot \bar{\bar{s}}_y \varepsilon_1$ and $\bar{\bar{\mu}}_2 = \bar{\bar{s}}_x \cdot \bar{\bar{s}}_y \mu_1$, then the two regions are perfectly matched. This can be further extended to the region where medium 2 also overlaps a PML layer with a z-directed normal. This is perfectly matched if $\bar{\bar{\varepsilon}}_3 = \bar{\bar{s}}_x \cdot \bar{\bar{s}}_y \cdot \bar{\bar{s}}_z \varepsilon_1$ and $\bar{\bar{\mu}}_3 = \bar{\bar{s}}_x \cdot \bar{\bar{s}}_y \cdot \bar{\bar{s}}_z \mu_1$.

In general, one can pose the PML layer as an anisotropic medium:

$$\bar{\bar{\varepsilon}} = \bar{\bar{s}}\varepsilon_h, \quad \bar{\bar{\mu}} = \bar{\bar{s}}\mu_h \tag{6.24}$$

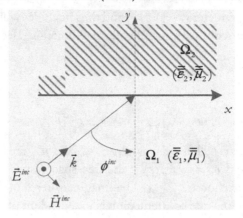

Figure 6.2: Uniform plane wave incident on a uniaxial media half-space with a y-normal.

where (ε_h, μ_h) is the material parameters of the host medium and

$$
\bar{\bar{s}} = \bar{\bar{s}}_x \cdot \bar{\bar{s}}_y \cdot \bar{\bar{s}}_z =
\begin{bmatrix}
\dfrac{s_y s_z}{s_x} & 0 & 0 \\
0 & \dfrac{s_x s_z}{s_y} & 0 \\
0 & 0 & \dfrac{s_x s_y}{s_z}
\end{bmatrix}
\tag{6.25}
$$

where $\bar{\bar{s}}$ is the anisotropic PML tensor. Inside the PML, s_k is defined via (6.20). In regions outside of the PML, $\kappa_k = 1$ and $\sigma_k = 0$, resulting in $s_k = 1$ ($k = x, y,$ or z). Thus, a single tensor of this form can be applied to any of the PML layers.

6.3   STRETCHED COORDINATE FORM OF THE PML

Chew and Weedon [4] showed that the PML can also be posed in a stretched coordinate frame with complex metric-tensor coefficients. The partial derivatives in the stretched coordinate space are expressed as:

$$
\frac{\partial}{\partial \tilde{x}} = \frac{1}{s_x}\frac{\partial}{\partial x}, \qquad
\frac{\partial}{\partial \tilde{y}} = \frac{1}{s_y}\frac{\partial}{\partial y}, \qquad
\text{and } \frac{\partial}{\partial \tilde{z}} = \frac{1}{s_z}\frac{\partial}{\partial z}.
\tag{6.26}
$$

The gradient operator is thus expressed in stretched coordinates as:

$$
\tilde{\nabla} = \hat{x}\frac{1}{s_x}\frac{\partial}{\partial x} + \hat{y}\frac{1}{s_y}\frac{\partial}{\partial y} + \hat{z}\frac{1}{s_z}\frac{\partial}{\partial z}.
\tag{6.27}
$$

Consequently, the time-harmonic Maxwell's curl equations are expressed in the complex-coordinate space as:

$$-j\omega\mu\vec{H} = \tilde{\nabla} \times \vec{E} = \hat{x}\left(\frac{1}{s_y}\frac{\partial}{\partial y}E_z - \frac{1}{s_z}\frac{\partial}{\partial z}E_y\right) + \hat{y}\left(\frac{1}{s_z}\frac{\partial}{\partial z}E_x - \frac{1}{s_x}\frac{\partial}{\partial x}E_z\right)$$

$$+ \hat{z}\left(\frac{1}{s_x}\frac{\partial}{\partial x}E_y - \frac{1}{s_y}\frac{\partial}{\partial y}E_x\right) \tag{6.28}$$

$$j\omega\varepsilon\vec{E} = \tilde{\nabla} \times \vec{H} = \hat{x}\left(\frac{1}{s_y}\frac{\partial}{\partial y}H_z - \frac{1}{s_z}\frac{\partial}{\partial z}H_y\right) + \hat{y}\left(\frac{1}{s_z}\frac{\partial}{\partial z}H_x - \frac{1}{s_x}\frac{\partial}{\partial x}H_z\right) +$$

$$+ \hat{z}\left(\frac{1}{s_x}\frac{\partial}{\partial x}H_y - \frac{1}{s_y}\frac{\partial}{\partial y}H_x\right). \tag{6.29}$$

Consider the half-space problem in Fig. 6.1. In this case, rather than posing Ω_2 to be an anisotropic medium, it is posed as a "stretched-coordinate" medium, with $s_y = s_z = 1$ (non-overlapping case), and s_x being defined by (6.20). Consider the TE$_z$-polarized case, with the fields in Ω_1 defined by (6.11). From, (6.28) and (6.29) it can be shown that the plane wave transmitted into Ω_2 can be expressed as:

$$\vec{H}_2 = \hat{z}\,H_0\,\tau\,e^{-jk_{2x}x - jk_{2y}y}, \tag{6.30}$$

$$\vec{E}_2 = \left(-\hat{x}\,\frac{k_{2y}}{\omega\,\varepsilon_2} + \hat{y}\,\frac{k_{2x}}{\omega\,\varepsilon_2 s_x}\right)H_0\,\tau\,e^{-jk_{2x}x - jk_{2y}y} \tag{6.31}$$

where the k-vector satisfies the dispersion relation:

$$k_{2x} = s_x\sqrt{k_2^2 - k_{1y}^2} \tag{6.32}$$

The reflection and transmission coefficients, Γ and τ, are found by enforcing continuity of the tangential electric and magnetic fields across the PML interface boundary (assumed to be located at $x = 0$ for simplicity). For the field continuity to be independent of y, there must be a phase matching: $k_{2y} = k_{1y}$. Then, enforcing the boundary condition, Γ and τ are found to be:

$$\Gamma = \frac{\frac{k_{1x}}{\varepsilon_1} - \frac{k_{2x}}{\varepsilon_2 s_x}}{\frac{k_{1x}}{\varepsilon_1} + \frac{k_{2x}}{\varepsilon_2 s_x}}, \qquad \tau = 1 + \Gamma = \frac{2\frac{k_{1x}}{\varepsilon_1}}{\frac{k_{1x}}{\varepsilon_1} + \frac{k_{2x}}{\varepsilon_2}}. \tag{6.33}$$

It is observed that, if $\varepsilon_2 = \varepsilon_1$ and $\mu_2 = \mu_1$, then $k_2 = k_1$, and from the dispersion relation (6.32),

$$k_{2x} = s_x k_{1x} \tag{6.34}$$

Consequently, $\Gamma \equiv 0$. A similar proof can be derived for the TM$_z$ polarized case. Therefore, this satisfies the condition for a perfectly matched medium.

From (6.30), it is observed that the wave transmitted into Ω_2 will have the propagation characteristic along the normal direction of the form:

$$e^{-jk_1 s_x x} = e^{-jk_{1x} \kappa_x x} e^{-\sqrt{\varepsilon_{r_1} \mu_{r_1}} \eta_o \sigma_x \cos \phi^{inc} x} \tag{6.35}$$

which is identical to that derived via the anisotropic PML formulation presented in Section 6.2.

The stretched coordinate PML shares the same properties as the anisotropic form. The first is that the interface is reflectionless. The second is that plane waves in the two PML media share identical dispersion relations. Consequently, both PML media have identical absorptive properties. The one distinction between the two representations is that for the anisotropic PML, the normal flux $s_x^{-1} \varepsilon E_x$ is continuous across the PML boundary interface. For the stretched coordinate form, the normal electric field E_x is continuous across the PML boundary interface since there is no actual discontinuity in the material parameter. In practice, either method has equivalent properties as an absorbing medium when applied to terminating FDTD grids.

6.4   PML REFLECTION ERROR

6.4.1   THE IDEAL PML

Thin layers of PML will be placed near the outer boundaries of the FDTD grid. The layers will have a finite thickness, d. The exterior boundary of the grid $\partial \Omega$ is simply assumed to be a PEC wall, as illustrated in Fig. 6.3. Due to the finite thickness, the wave that penetrates into the PML

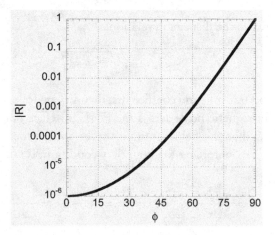

Figure 6.3: Uniform plane wave incident on a uniaxial media half-space with a y-normal.

will reflect off the back wall, and thus return into the working volume without reflection once again. As a result, the constitutive parameters of the PML, and the thickness, should be set such that the signal is sufficiently attenuated prior to it returning back into the problem domain.

From (6.21) or (6.35), it can be shown that the effective reflection error of the PEC-backed PML has the magnitude:

$$R\left(\phi\right) = e^{-2\sqrt{\varepsilon_{r_1}\mu_{r_1}}\eta_o\sigma_x\cos\phi d} \tag{6.36}$$

where the factor of 2 accounts for the round trip attenuation through the PML. The PML reflection error is minimum for $\phi = 0$, and $R(90°) = 1$ at grazing incidence. Typically, for a fixed thickness d, σ_x will be chosen to minimize error for normal incidence. For example, the graph in Fig. 6.3 illustrates the reflection error versus ϕ when $R\left(0°\right) = 10^{-6}$. It is noted that the angular spread over which the reflection error is below 1% reflection error is much broader than that of a local ABC (c.f., Fig. 5.4). Though, at high angles of incidence the reflection error does get large. Nevertheless, this is typically of little concern if such reflected waves are near corners where they will reflect into an adjacent PML layer near normal incidence, as opposed to being reflected back towards the device under test.

6.4.2   PML PARAMETER SCALING

Ideally, the conductivity of the PML can be made significantly large such that the reflection error is driven to nearly zero. Unfortunately, there is a practical limit to how large the conductivity can be when the PML is employed within discrete space such as the FDTD method. One reason for this is the discrete fields are approximated via a piecewise constant interpolation. As a result, if the fields decay too rapidly, they cannot be accurately resolved. Secondly, the Yee-discretization is based on a staggering of the electric and magnetic fields in space. As a consequence, the electric and magnetic materials are also staggered in space (c.f., Section 3.7). However, the PML relies on a distinct duality of the electric and magnetic properties of the material (e.g., (6.24)). Due to the staggering of the materials, the first one-half of a cell of the PML layer will only be the electric material. As a consequence, this thin layer is not perfectly matched. This can lead to a significant reflection off the front of the PML.

In order to compensate for such large reflection errors, Bérenger astutely recommended that the constitutive parameters be scaled. This will significantly reduce the reflection off the primal interface, allowing the wave to enter into the material and hence be absorbed.

The PML constitutive parameters can be expressed as one-dimensional functions without violating Maxwell's equations. That is, we define the one-dimensional functions: $s_x(x)$, $s_y(y)$, and $s_z(z)$, which are still defined by (6.20), where σ_x and κ_x are now one-dimensional functions of x. An appropriate scaling function is a polynomial scaling such that:

$$\sigma_x\left(x\right) = \begin{cases} \frac{|x-x_o|^m}{d^m}\sigma_x^{\max}, & x_o \le x \le x_o + d \\ 0, & \text{else} \end{cases} \tag{6.37}$$

and

$$\kappa_x\left(x\right) = \begin{cases} 1 + \frac{|x-x_o|^m}{d^m}\left(\kappa_x^{\max} - 1\right), & x_o \le x \le x_o + d \\ 1, & \text{else} \end{cases} \tag{6.38}$$

where x_o is the x-coordinate of the PML interface, d is the thickness of the PML layer, and $\sigma_x^{\max}$ is the maximum value of σ_x at $x = x_o + d$. Similarly, $\kappa_x^{\max}$ is the maximum value of κ_x at $x = x_o + d$. Similar scaling functions are used for $s_y(y)$, and $s_z(z)$.

Note that the PML is still perfectly matched when the constitutive parameters are scaled along the normal direction. This is easily shown in stretched coordinates. Consider the PML layer to be broken up into N thin layers, each of thickness Δ_x. Each individual layer is perfectly matched. The attenuation of the wave is accumulated, such that, from (6.36), the total reflection of the layered PML is:

$$
R(\phi) = e^{-2\sqrt{\varepsilon_r \mu_r}\eta_o \cos\phi\sigma_x(x_o)\Delta_x} \cdot e^{-2\sqrt{\varepsilon_r \mu_r}\eta_o \cos\phi\sigma_x(x_o+\Delta_x)\Delta_x} \ldots e^{-2\sqrt{\varepsilon_r \mu_r}\eta_o \cos\phi\sigma_x(x_o+(N-1)\Delta_x)\Delta_x}
$$
$$
= e^{-2\sqrt{\varepsilon_r \mu_r}\eta_o \cos\phi\Delta_x \sum_{i=0}^{N-1} \sigma_x(x_o+i\Delta_x)}.
$$

$$(6.39)$$

In the limit that $\Delta_x \to 0$ (and $N \to \infty$):

$$
R(\phi) = e^{-2\sqrt{\varepsilon_r \mu_r}\eta_o \cos\phi \int_{x=x_o}^{x_o+d} \sigma_x(x)dx}.
$$

$$(6.40)$$

For the polynomial scaled parameters in (6.37), the reflection coefficient of a propagating plane wave is expressed in closed form as:

$$
R(\phi) = e^{-2\sqrt{\varepsilon_r \mu_r}\eta_o \cos\phi\sigma_x^{\max}d/(m+1)}.
$$

$$(6.41)$$

Typically, when designing a PML media for a FDTD application, d is fixed, and one chooses m and then determines $\sigma_x^{\max}$. This is typically done by specifying a value for $R(0)$. Then, from (6.41),

$$
\sigma_x^{\max} = -\frac{(m+1)\ln[R(0)]}{2\eta_o d\sqrt{\varepsilon_r \mu_r}}.
$$

$$(6.42)$$

If $\sigma_x^{\max}$ is sufficiently small, then the reflection error is dominated by the reflection off the terminating PEC boundary, as predicted by (6.41). If the $\sigma_x^{\max}$ is too large, then the reflection error is dominated by discretization error. This error tends to grow with $\sigma_x^{\max}$. At the transition between these two regions, there is an optimal value of $\sigma_x^{\max}$. A heuristic formula was proposed in [6, 9] that provides a close estimate for the optimal value of $\sigma_x^{\max}$. That is, the optimal $\sigma^{\max}$ value is:

$$
\sigma_x^{\text{opt}} = \frac{0.8(m+1)}{\eta_o \Delta x\sqrt{\varepsilon_{r,eff}\,\mu_{r,eff}}},
$$

$$(6.43)$$

where Δx is the cell spacing of the FDTD grid along the x-direction. Also, note that $\varepsilon_{r,eff}$ and $\mu_{r,eff}$ are the "effective" relative permittivity and permeability of the boundary. If the local material media is inhomogeneous, then these can be averaged values of the material parameters.

Equation (6.43) is derived based on the observation that the optimal reflection error for a 10-cell thick (i.e., $d = 10\Delta x$) polynomial scaled PML is approximately $R(0) = e^{-16}$. Also, the optimal reflection error for a 5-cell thick polynomial scaled PML is approximately $R(0) = e^{-8}$. With this information, σ_x^{opt} is derived directly from (6.42).

Finally, it has also been found that the optimal order for the polynomial scaling is typically on the range of $3 \leq m \leq 4$.

6.4.3   REFLECTION ERROR

It was discussed in the previous section that the PML suffers from significant reflection error off the front boundary interface due to the staggering of the FDTD grid. This was resolved by scaling the PML parameters. Nevertheless, even with scaling, it is found that large reflections can still occur at low frequencies. To demonstrate this, consider a TE_z-polarized plane wave that is normally incident on the PML boundary. From (6.14), the reflection coefficient off the boundary is:

$$\Gamma = \frac{1 - \sqrt{d/b}}{1 + \sqrt{d/b}}, \tag{6.44}$$

where it has been assumed that $\varepsilon_1 = \varepsilon_2$, $\mu_1 = \mu_2$, and from (6.15), $k_{2_x} = k_1\sqrt{b}$. (b and d are defined in (6.4)). Note that we have assumed $b \neq d$ since due to staggering of the grid, the electric and magnetic materials are not equal in the first one-half cell thick layer of the PML region. For a simpler notation, we can express

$$b = s_x(x), \quad d = s_x^*(x) \tag{6.45}$$

where s_x is associated with the electric material, and s_x^* is associated with the magnetic material.

It is assumed that both s_x and s_x^* are scaled, as in (6.37) and (6.38). Thus, at the front PML boundary, both $s_x(x_o) = s_x^*(x_o) = 1$. The next boundary occurs at $x = x_o + \Delta_x/2$. At this boundary, $s_x^*(x_o + \Delta_x/2) = 1$, due to the staggering of the grid. However, $s_x(x_o + \Delta_x/2) \neq 1$. Therefore,

$$\Gamma_o = \frac{\sqrt{s_x(x_o + \Delta_x/2)} - 1}{\sqrt{s_x(x_o + \Delta_x/2)} + 1}. \tag{6.46}$$

For simplicity, it is assumed that $\kappa_x = 1$. Thefore:

$$\Gamma_o = \frac{\sqrt{1 + \sigma_x(x_o + \Delta_x/2)/j\omega\varepsilon_o} - 1}{\sqrt{1 + \sigma_x(x_o + \Delta_x/2)/j\omega\varepsilon_o} + 1}. \tag{6.47}$$

For frequencies when $\omega >> \sigma_x(x_o + \Delta_x/2)/\varepsilon_o$, $\Gamma_o = 0$. However, at low frequencies, when $\omega << \sigma_x(x_o + \Delta_x/2)/\varepsilon_o$, $\Gamma_o = 1$. Consequently, low frequencies can suffer a very large reflection error off the front PML boundary. The break frequency between these two regions is defined as:

$$f_c = \sigma_x(x_o + \Delta_x/2)/2\pi\varepsilon_o = 1.8 \times 10^{10}\sigma_x(x_o + \Delta_x/2) \text{ Hz.} \tag{6.48}$$

Therefore, when $f > f_c$, the reflection off the front PML boundary is assumed to be quite small. Else, when $f > f_c$, the reflection off the front PML boundary is assumed to be quite large.

When simulating time-dependent fields, low frequency correlates to waves having a long interaction time with the PML boundary. This could include slow waves or slowly decaying near-fields. Thus, an interaction time can also be defined as:

$$t_c = \frac{1}{f_c} = \frac{5.56 \times 10^{-11}}{\sigma_x (x_o + \Delta_x/2)} \ \text{s.} \tag{6.49}$$

Waves interacting longer than t_c with the PML boundary will suffer from large reflections off the front boundary.

6.4.4 THE COMPLEX FREQUENCY SHIFTED (CFS) PML

The low-frequency reflection error off the PML boundary can be reduced by decreasing $\sigma_x (x_o + \Delta_x/2)$, by either reducing σ_x^{max}, at the expense of lower absorption, or by increasing d, at the expense of a larger solution space. Nevertheless, since f_c is always > 0, this can never be sufficient. A remedy is to choose an alternate form for s_x. The preferred choice is what is referred to as the Complex-Frequency Shifted (CFS) tensor coefficient [7, 10]:

$$s_x = \kappa_x + \frac{\sigma_x}{a_x + j\omega\varepsilon_o} \tag{6.50}$$

where a_x is a non-negative real number. The CFS-tensor coefficient is so called since it effectively shifts the pole of s_x off the origin and into the lower complex plane.

Using the CFS-tensor coefficient, the reflection coefficient due to the front boundary interface as derived in (6.46) is:

$$\Gamma_o = \frac{\sqrt{1 + \sigma_x(x_o + \Delta_x/2)/(a_x(x_o + \Delta_x/2) + j\omega\varepsilon_o)} - 1}{\sqrt{1 + \sigma_x(x_o + \Delta_x/2)/(a_x(x_o + \Delta_x/2) + j\omega\varepsilon_o)} + 1}. \tag{6.51}$$

At high frequencies, where $\omega >> \sigma_x(x_o + \Delta_x/2)/\varepsilon_o$, $\Gamma_o = 0$ as expected. Then, consider the low frequency limit where $\omega << a_x(x_o + \Delta_x/2)/\varepsilon_o$:

$$\Gamma_o = \frac{\sqrt{1 + \sigma_x(x_o + \Delta_x/2)/a_x(x_o + \Delta_x/2)} - 1}{\sqrt{1 + \sigma_x(x_o + \Delta_x/2)/a_x(x_o + \Delta_x/2)} + 1}. \tag{6.52}$$

Next, consider what happens if $\sigma_x(x_o + \Delta_x/2)/a_x(x_o + \Delta_x/2) << 1$. Using a Taylor series expansion, $\sqrt{1+x} \approx 1 + \frac{1}{2}x$, when $x << 1$. As a result, (6.52) becomes:

$$\Gamma_o \simeq \sigma_x(x_o + \Delta_x/2)/4a_x(x_o + \Delta_x/2). \tag{6.53}$$

Consequently, the CFS tensor coefficient has the ability to mitigate the low-frequency issues that can plague the PML.

The next question to be addressed is how will the CFS-PML impact the attenuation of a wave propagating through the PML medium. Assume that the wave entering the PML has a complex wavenumber $\gamma_x = \alpha_x + jk_x$. Consequently, once in the PML, the wave will have the propagation term:

$$e^{-(\alpha_x+jk_x)(\kappa_x+\frac{\sigma_x}{a_x+j\omega\varepsilon_0})x} = e^{-(\alpha_x\kappa_x+jk_x\kappa_x+\alpha_x\frac{\sigma_x}{a_x+j\omega\varepsilon_0}+jk_x\frac{\sigma_x}{a_x+j\omega\varepsilon_0})}. \tag{6.54}$$

It is observed that κ_x will amplify the attenuation of the original wave. Next, at high frequencies, $\omega >> a_x/\varepsilon_0$, the CFS-PML will attenuate the propagating wave in the same manner as the original PML. However, at low frequencies, $\omega << a_x/\varepsilon_0$, it is observed that the propagating wave will not attenuate at all! This break frequency occurs at:

$$f_a = a_x/2\pi\varepsilon_0. \tag{6.55}$$

Thus, if $f << f_a$, propagating waves will not be attenuated. Otherwise, when $f >> f_a$, propagating waves are highly attenuated.

In summary, based on (6.53), $a_x(x_o + \Delta_x/2)$ must be much larger than $\sigma_x(x_o + \Delta_x/2)$ to eliminate low frequency reflection errors at the front boundary interface. On the other hand, a_x must be small within the PML in order to improve the attenuation of low-frequency propagating waves within the PML. Therefore, it is concluded that a_x should be scaled as a function of x. However, it should be scaled in a manner such that it is *maximum* at x_o, and minimum at the terminating boundary at $x_o + d$. This is opposite of how σ_x and κ_x were scaled. Using polynomial scaling,

$$a_x(x) = \begin{cases} a_x^{\max}\left|\frac{d+x_o-x}{d}\right|^m, & x_o \le x \le x_o + d \\ 0, & \text{else}. \end{cases} \tag{6.56}$$

Meanwhile, σ_x and κ_x are still scaled as specified by (6.37) and (6.38).

6.5   IMPLEMENTING THE CFS-PML IN THE FDTD METHOD

To date, the CFS-PML method is the most robust method for terminating an FDTD lattice [3, 11]. It has the advantage that it absorbs both evanescent and propagating waves at high or low frequencies [12, 13]. It also avoids late time instabilities [14]. The most efficient implementation of the CFS-PML is based on the stretched coordinate formulation [7, 15], which is presented here.

6.5.1   AN ADE FORM OF THE CFS-PML

Consider, the x-projection of Faraday's and Ampere's laws as expressed in stretched coordinates in (6.28) and (6.29), respectively:

$$-j\omega\mu H_x = \frac{1}{s_y}\frac{\partial}{\partial y}E_z - \frac{1}{s_z}\frac{\partial}{\partial z}E_y \tag{6.57}$$

$$j\omega\varepsilon E_x = \frac{1}{s_y}\frac{\partial}{\partial y}H_z - \frac{1}{s_z}\frac{\partial}{\partial z}H_y \tag{6.58}$$

where a full overlap in three-dimensions is assumed. The PML tensor coefficients are assumed to be the CFS-PML coefficients of (6.50). Transforming (6.57) or (6.58) into the time domain leads to a convolution between the inverse of the stretched coordinate parameters and the partial derivatives [7]. Alternatively, an auxiliary variable can be introduced which is constrained via the introduction of the appropriate ADE based on the identity:

$$\frac{1}{s_y} = \frac{1}{\kappa_y} - \frac{1}{B_y} \tag{6.59}$$

where

$$\frac{1}{B_y} = \frac{\sigma_y}{\kappa_y} \frac{1}{\left(\kappa_y \left(j\omega\varepsilon_o + a_y\right) + \sigma_y\right)}. \tag{6.60}$$

Therefore, one can express

$$\frac{1}{s_y} \frac{\partial}{\partial y} E_z = \frac{1}{\kappa_y} \frac{\partial}{\partial y} E_z + Q_{y,z}^E. \tag{6.61}$$

where the new auxiliary variable $Q_{y,z}^E$ satisfies:

$$Q_{y,z}^E = -\frac{1}{B_y} \frac{\partial}{\partial y} E_z. \tag{6.62}$$

Substituting (6.60) for $1/B_y$, this expression is also is transformed into the time-domain, leading to the Auxiliary Differential Equation (ADE):

$$\kappa_y \varepsilon_o \frac{d}{dt} Q_{y,z}^E + \left(\kappa_y a_y + \sigma_y\right) Q_{y,z}^E = -\frac{\sigma_y}{\kappa_y} \frac{\partial}{\partial y} E_z \tag{6.63}$$

which is an ordinary differential equation in time. Similarly, another auxiliary variable $Q_{z,y}^E$ is introduced, which satisfies the ADE:

$$\kappa_z \varepsilon_o \frac{d}{dt} Q_{z,y}^E + \left(\kappa_z a_z + \sigma_z\right) Q_{z,y}^E = -\frac{\sigma_z}{\kappa_z} \frac{\partial}{\partial z} E_y. \tag{6.64}$$

In conclusion, (6.57) and (6.58) can be rewritten as:

$$-\mu \frac{\partial}{\partial t} H_x = \frac{1}{\kappa_y} \frac{\partial}{\partial y} E_z - \frac{1}{\kappa_z} \frac{\partial}{\partial z} E_y + Q_{y,z}^E - Q_{z,y}^E \tag{6.65}$$

and,

$$\varepsilon \frac{\partial}{\partial t} E_x = \frac{1}{\kappa_y} \frac{\partial}{\partial y} H_z - \frac{1}{\kappa_z} \frac{\partial}{\partial z} H_y + Q_{y,z}^H - Q_{z,y}^H \tag{6.66}$$

where the auxiliary variables $Q_{y,z}^E$ and $Q_{z,y}^E$ satisfy the ADEs (6.63) and (6.64), respectively. Similarly, $Q_{y,z}^H$ and $Q_{z,y}^H$ satisfy:

$$\kappa_y \varepsilon_o \frac{d}{dt} Q_{y,z}^H + \left(\kappa_y a_y + \sigma_y\right) Q_{y,z}^H = -\frac{\sigma_y}{\kappa_y} \frac{\partial}{\partial y} H_z \tag{6.67}$$

$$\kappa_z \varepsilon_o \frac{d}{dt} Q_{z,y}^H + \left(\kappa_z a_z + \sigma_z\right) Q_{z,y}^H = -\frac{\sigma_z}{\kappa_z} \frac{\partial}{\partial z} H_y. \tag{6.68}$$

From the y- and z-projections of Faraday's and Ampere's laws:

$$-\mu\frac{\partial}{\partial t}H_y = \frac{1}{\kappa_z}\frac{\partial}{\partial z}E_x - \frac{1}{\kappa_x}\frac{\partial}{\partial x}E_z + Q^E_{z,x} - Q^E_{x,z} \tag{6.69}$$

$$-\mu\frac{\partial}{\partial t}H_z = \frac{1}{\kappa_x}\frac{\partial}{\partial x}E_y - \frac{1}{\kappa_y}\frac{\partial}{\partial y}E_x + Q^E_{x,y} - Q^E_{y,x} \tag{6.70}$$

$$\varepsilon\frac{\partial}{\partial t}E_y = \frac{1}{\kappa_z}\frac{\partial}{\partial z}H_x - \frac{1}{\kappa_x}\frac{\partial}{\partial x}H_z + Q^H_{z,x} - Q^H_{x,z} \tag{6.71}$$

and

$$\varepsilon\frac{\partial}{\partial t}E_z = \frac{1}{\kappa_x}\frac{\partial}{\partial x}H_y - \frac{1}{\kappa_y}\frac{\partial}{\partial y}H_x + Q^H_{x,y} - Q^H_{y,x} \tag{6.72}$$

where the auxiliary variables satisfy the ADEs of the form

$$\kappa_i\varepsilon_o\frac{d}{dt}Q^H_{i,j} + (\kappa_i a_i + \sigma_i)Q^H_{i,j} = -\frac{\sigma_i}{\kappa_i}\frac{\partial}{\partial i}H_j \tag{6.73}$$

and

$$\kappa_i\varepsilon_o\frac{d}{dt}Q^E_{i,j} + (\kappa_i a_i + \sigma_i)Q^E_{i,j} = -\frac{\sigma_i}{\kappa_i}\frac{\partial}{\partial i}E_j \tag{6.74}$$

where i and $j = x, y, z$ ($i \neq j$). This constitutes the ADE CFS-PML algorithm.

6.5.2 YEE-ALGORITHM FOR THE CFS-PML

The CFS-PML algorithm introduces a set of new auxiliary variables which must be discretized in a manner that is consistent with the underlying Yee-algorithm. Starting with $Q^E_{y,z}$, it is seen that $Q^E_{y,z}$ is a function of the y-derivative of E_z (c.f., (6.62)). Thus, $Q^E_{y,z}$ should be sampled at the same discrete time as E_z. This term is also used to evaluate H_x (6.65). Therefore, the discrete value of $Q^E_{y,z}$ should be sampled at the same spatial location as H_x. Following this strategy, and the basic Yee-discretization presented in Chapter 3, the explicit update equations derived from Faraday's and Ampere's laws in a PML region terminating a lossless, isotropic medium are expressed as:

$$H^{n+1}_{x\,i,j+\frac{1}{2},k+\frac{1}{2}} = H^{n}_{x\,i,j+\frac{1}{2},k+\frac{1}{2}} +$$
$$\frac{\Delta t}{\mu}\left[\left(\frac{E^{n+\frac{1}{2}}_{y\,i,j+\frac{1}{2},k+1} - E^{n+\frac{1}{2}}_{y\,i,j+\frac{1}{2},k}}{\kappa_z\Delta z}\right) - \left(\frac{E^{n+\frac{1}{2}}_{z\,i,j+1,k+\frac{1}{2}} - E^{n+\frac{1}{2}}_{z\,i,j,k+\frac{1}{2}}}{\kappa_y\Delta y}\right) + Q^{E^{n+\frac{1}{2}}}_{z,y\,i,j+\frac{1}{2},k+\frac{1}{2}} - Q^{E^{n+\frac{1}{2}}}_{y,z\,i,j+\frac{1}{2},k+\frac{1}{2}}\right] \tag{6.75}$$

$$H^{n+1}_{y\,i+\frac{1}{2},j,k+\frac{1}{2}} = H^{n}_{y\,i+\frac{1}{2},j,k+\frac{1}{2}} +$$
$$\frac{\Delta t}{\mu}\left[\left(\frac{E^{n+\frac{1}{2}}_{z\,i+1,j,k+\frac{1}{2}} - E^{n+\frac{1}{2}}_{z\,i,j,k+\frac{1}{2}}}{\kappa_x\Delta x}\right) - \left(\frac{E^{n+\frac{1}{2}}_{x\,i+\frac{1}{2},j,k+1} - E^{n+\frac{1}{2}}_{x\,i+\frac{1}{2},j,k}}{\kappa_z\Delta z}\right) + Q^{E^{n+\frac{1}{2}}}_{x,z\,i+\frac{1}{2},j,k+\frac{1}{2}} - Q^{E^{n+\frac{1}{2}}}_{z,x\,i+\frac{1}{2},j,k+\frac{1}{2}}\right] \tag{6.76}$$

$$H_{z}{}^{n+1}_{i+\frac{1}{2},j+\frac{1}{2},k} = H_{z}{}^{n}_{i+\frac{1}{2},j+\frac{1}{2},k} + \frac{\Delta t}{\mu}\left[\left(\frac{E_{x}{}^{n+\frac{1}{2}}_{i+\frac{1}{2},j+1,k} - E_{x}{}^{n+\frac{1}{2}}_{i+\frac{1}{2},j,k}}{\kappa_{y}\Delta y}\right) - \left(\frac{E_{y}{}^{n+\frac{1}{2}}_{i+1,j+\frac{1}{2},k} - E_{y}{}^{n+\frac{1}{2}}_{i,j+\frac{1}{2},k}}{\kappa_{x}\Delta x}\right) + Q^{E}_{y,x}{}^{n+\frac{1}{2}}_{i+\frac{1}{2},j+\frac{1}{2},k} - Q^{E}_{x,y}{}^{n+\frac{1}{2}}_{i+\frac{1}{2},j+\frac{1}{2},k}\right] \tag{6.77}$$

$$E_{x}{}^{n+\frac{1}{2}}_{i+\frac{1}{2},j,k} = E_{x}{}^{n-\frac{1}{2}}_{i+\frac{1}{2},j,k} + \frac{\Delta t}{\varepsilon}\left[\left(\frac{H_{z}{}^{n}_{i+\frac{1}{2},j+\frac{1}{2},k} - H_{z}{}^{n}_{i+\frac{1}{2},j-\frac{1}{2},k}}{\kappa_{y}\Delta y}\right) - \left(\frac{H_{y}{}^{n}_{i+\frac{1}{2},j,k+\frac{1}{2}} - H_{y}{}^{n}_{i+\frac{1}{2},j,k-\frac{1}{2}}}{\kappa_{z}\Delta z}\right) + Q^{H}_{y,z}{}^{n}_{i+\frac{1}{2},j,k} - Q^{H}_{z,y}{}^{n}_{i+\frac{1}{2},j,k}\right] \tag{6.78}$$

$$E_{y}{}^{n+\frac{1}{2}}_{i,j+\frac{1}{2},k} = E_{y}{}^{n-\frac{1}{2}}_{i,j+\frac{1}{2},k} + \frac{\Delta t}{\varepsilon}\left[\left(\frac{H_{x}{}^{n}_{i,j+\frac{1}{2},k+\frac{1}{2}} - H_{x}{}^{n}_{i,j+\frac{1}{2},k-\frac{1}{2}}}{\kappa_{z}\Delta z}\right) - \left(\frac{H_{z}{}^{n}_{i+\frac{1}{2},j+\frac{1}{2},k} - H_{z}{}^{n}_{i-\frac{1}{2},j+\frac{1}{2},k}}{\kappa_{x}\Delta x}\right) + Q^{H}_{z,x}{}^{n}_{i,j+\frac{1}{2},k} - Q^{H}_{x,z}{}^{n}_{i,j+\frac{1}{2},k}\right] \tag{6.79}$$

and

$$E_{z}{}^{n+\frac{1}{2}}_{i,j,k+\frac{1}{2}} = E_{z}{}^{n-\frac{1}{2}}_{i,j,k+\frac{1}{2}} + \frac{\Delta t}{\varepsilon}\left[\left(\frac{H_{y}{}^{n}_{i+\frac{1}{2},j,k+\frac{1}{2}} - H_{y}{}^{n}_{i-\frac{1}{2},j,k+\frac{1}{2}}}{\kappa_{x}\Delta x}\right) - \left(\frac{H_{x}{}^{n}_{i,j+\frac{1}{2},k+\frac{1}{2}} - H_{x}{}^{n}_{i,j-\frac{1}{2},k+\frac{1}{2}}}{\kappa_{y}\Delta y}\right) + Q^{H}_{x,y}{}^{n}_{i,j,k+\frac{1}{2}} - Q^{H}_{y,x}{}^{n}_{i,j,k+\frac{1}{2}}\right]. \tag{6.80}$$

The last step is to derive solutions to the ADE equations. As an example, consider the solution of Equation (6.63). This equation is a first-order ordinary-differential equation for $Q^{E}_{y,z}$ driven by the forcing function $-\frac{\sigma_{y}}{\kappa_{y}}\frac{\partial}{\partial y}E_{z}$. For second-order accuracy, it is sufficient to approximate the forcing function as a piecewise constant function. Then, each time interval $(n - \frac{1}{2})\Delta t < t < (n + \frac{1}{2})\Delta t$ can be represented via a unit-step response solution, with source $\frac{\partial}{\partial y}E_{z}^{n+\frac{1}{2}}u(t - (n - \frac{1}{2})\Delta t)$, where $u(t)$ is the unit-step function, and initial condition $Q^{E}_{y,z}((n - \frac{1}{2})\Delta t) = Q^{E}_{y,z}{}^{n-\frac{1}{2}}$ from the previous time interval. From (6.63), it is seen that $Q^{E}_{y,z}$ has the steady-state value:

$$Q^{E}_{y,z}(t \to \infty) = -\frac{\sigma_{y}}{\kappa_{y}}\frac{1}{(\kappa_{y}a_{y} + \sigma_{y})}\frac{\partial E_{z}^{n}}{\partial y} \tag{6.81}$$

The unit-step response to (6.63) is then expressed as:

$$Q^{E}_{y,z}(t) = \left(Q^{E}_{y,z}((n - \frac{1}{2})\Delta t) - Q^{E}_{y,z}(\infty)\right)e^{-\left(\frac{t-(n-\frac{1}{2})\Delta t}{\tau}\right)} + Q^{E}_{y,z}(\infty) \tag{6.82}$$

where,

$$\tau = \frac{\kappa_y \varepsilon_o}{\left(\kappa_y a_y + \sigma_y\right)} \tag{6.83}$$

is the time-constant. Finally, letting $Q_{y,z}^{E^{n+\frac{1}{2}}} = Q_{y,z}^E((n+\frac{1}{2})\Delta t)$, then from (6.82):

$$Q_{y,z}^{E^{n+\frac{1}{2}}} = b_y Q_{y,z}^{E^{n-\frac{1}{2}}} - c_y \frac{\partial E_z^{n+\frac{1}{2}}}{\partial y} \tag{6.84}$$

where,

$$b_y = e^{-\Delta t/\tau}, \qquad c_y = \frac{\sigma_y}{\kappa_y} \frac{1}{\left(\kappa_y a_y + \sigma_y\right)} \left(1 - e^{-\Delta t/\tau}\right) \tag{6.85}$$

and, τ is given by (6.83).

Comparing the recursive update in (6.84) to Roden's CPML formulation [7], it is observed that the expression is *identical* to that derived using the recursive convolution method. Consequently, the two methods are equivalent. The present method is somewhat easier to follow.

Finally, the update equations for the auxiliary variables are summarized as:

$$Q_{y,z}^{E^{n+\frac{1}{2}}}\Big|_{i,j+\frac{1}{2},k+\frac{1}{2}} = b_{y_{j+\frac{1}{2}}} Q_{y,z}^{E^{n-\frac{1}{2}}}\Big|_{i,j+\frac{1}{2},k+\frac{1}{2}} - c_{y_{j+\frac{1}{2}}} \left(\frac{E_z^{n+\frac{1}{2}}\Big|_{i,j+1,k+\frac{1}{2}} - E_z^{n+\frac{1}{2}}\Big|_{i,j,k+\frac{1}{2}}}{\Delta y} \right), \tag{6.86}$$

$$Q_{z,y}^{E^{n+\frac{1}{2}}}\Big|_{i,j+\frac{1}{2},k+\frac{1}{2}} = b_{z_{k+\frac{1}{2}}} Q_{z,y}^{E^{n-\frac{1}{2}}}\Big|_{i,j+\frac{1}{2},k+\frac{1}{2}} - c_{z_{k+\frac{1}{2}}} \left(\frac{E_y^{n+\frac{1}{2}}\Big|_{i,j+\frac{1}{2},k+1} - E_y^{n+\frac{1}{2}}\Big|_{i,j+\frac{1}{2},k}}{\Delta z} \right), \tag{6.87}$$

$$Q_{z,x}^{E^{n+\frac{1}{2}}}\Big|_{i+\frac{1}{2},j,k+\frac{1}{2}} = b_{z_{k+\frac{1}{2}}} Q_{z,x}^{E^{n-\frac{1}{2}}}\Big|_{i+\frac{1}{2},j,k+\frac{1}{2}} - c_{z_{k+\frac{1}{2}}} \left(\frac{E_x^{n+\frac{1}{2}}\Big|_{i+\frac{1}{2},j,k+1} - E_x^{n+\frac{1}{2}}\Big|_{i+\frac{1}{2},j,k}}{\Delta z} \right), \tag{6.88}$$

$$Q_{x,z}^{E^{n+\frac{1}{2}}}\Big|_{i+\frac{1}{2},j,k+\frac{1}{2}} = b_{x_{i+\frac{1}{2}}} Q_{x,z}^{E^{n-\frac{1}{2}}}\Big|_{i+\frac{1}{2},j,k+\frac{1}{2}} - c_{x_{i+\frac{1}{2}}} \left(\frac{E_z^{n+\frac{1}{2}}\Big|_{i+1,j,k+\frac{1}{2}} - E_z^{n+\frac{1}{2}}\Big|_{i,j,k+\frac{1}{2}}}{\Delta x} \right), \tag{6.89}$$

$$Q^{E^{n+\frac{1}{2}}}_{x,y}{}_{i+\frac{1}{2},j+\frac{1}{2},k} = b_{x}{}_{i+\frac{1}{2}} Q^{E^{n-\frac{1}{2}}}_{x,y}{}_{i+\frac{1}{2},j+\frac{1}{2},k} - c_{x}{}_{i+\frac{1}{2}} \left(\frac{E^{n+\frac{1}{2}}_{y}{}_{i+1,j+\frac{1}{2},k} - E^{n+\frac{1}{2}}_{y}{}_{i,j+\frac{1}{2},k}}{\Delta x} \right), \quad (6.90)$$

$$Q^{E^{n+\frac{1}{2}}}_{y,x}{}_{i+\frac{1}{2},j+\frac{1}{2},k} = b_{y}{}_{j+\frac{1}{2}} Q^{E^{n-\frac{1}{2}}}_{y,x}{}_{i+\frac{1}{2},j+\frac{1}{2},k} - c_{y}{}_{j+\frac{1}{2}} \left(\frac{E^{n+\frac{1}{2}}_{x}{}_{i+\frac{1}{2},j+1,k} - E^{n+\frac{1}{2}}_{x}{}_{i+\frac{1}{2},j,k}}{\Delta y} \right), \quad (6.91)$$

$$Q^{H^{n}}_{y,z}{}_{i+\frac{1}{2},j,k} = b_{y_{j}} Q^{H^{n-1}}_{y,z}{}_{i+\frac{1}{2},j,k} - c_{y_{j}} \left(\frac{H^{n}_{z}{}_{i+\frac{1}{2},j+\frac{1}{2},k} - H^{n}_{z}{}_{i+\frac{1}{2},j-\frac{1}{2},k}}{\Delta y} \right), \quad (6.92)$$

$$Q^{H^{n}}_{z,y}{}_{i+\frac{1}{2},j,k} = b_{z_{k}} Q^{H^{n-1}}_{z,y}{}_{i+\frac{1}{2},j,k} - c_{z_{k}} \left(\frac{H^{n}_{y}{}_{i+\frac{1}{2},j,k+\frac{1}{2}} - H^{n}_{y}{}_{i+\frac{1}{2},j,k-\frac{1}{2}}}{\Delta z} \right), \quad (6.93)$$

$$Q^{H^{n}}_{z,x}{}_{i,j+\frac{1}{2},k} = b_{z_{k}} Q^{H^{n-1}}_{z,x}{}_{i,j+\frac{1}{2},k} - c_{z_{k}} \left(\frac{H^{n}_{x}{}_{i,j+\frac{1}{2},k+\frac{1}{2}} - H^{n}_{x}{}_{i,j+\frac{1}{2},k-\frac{1}{2}}}{\Delta z} \right), \quad (6.94)$$

$$Q^{H^{n}}_{x,z}{}_{i,j+\frac{1}{2},k} = b_{x_{i}} Q^{H^{n-1}}_{x,z}{}_{i,j+\frac{1}{2},k} - c_{x_{i}} \left(\frac{H^{n}_{z}{}_{i+\frac{1}{2},j+\frac{1}{2},k} - H^{n}_{z}{}_{i-\frac{1}{2},j+\frac{1}{2},k}}{\Delta x} \right), \quad (6.95)$$

$$Q^{H^{n}}_{x,y}{}_{i,j,k+\frac{1}{2}} = b_{x_{i}} Q^{H^{n-1}}_{x,y}{}_{i,j,k+\frac{1}{2}} - c_{x_{i}} \left(\frac{H^{n}_{y}{}_{i+\frac{1}{2},j,k+\frac{1}{2}} - H^{n}_{y}{}_{i-\frac{1}{2},j,k+\frac{1}{2}}}{\Delta x} \right), \quad (6.96)$$

and

$$Q^{H^{n}}_{y,x}{}_{i,j,k+\frac{1}{2}} = b_{y_{j}} Q^{H^{n-1}}_{y,x}{}_{i,j,k+\frac{1}{2}} - c_{y_{j}} \left(\frac{H^{n}_{x}{}_{i,j+\frac{1}{2},k+\frac{1}{2}} - H^{n}_{x}{}_{i,j-\frac{1}{2},k+\frac{1}{2}}}{\Delta y} \right). \quad (6.97)$$

It is noted that in (6.86)–(6.97), the coefficients b and c have subscripts indicating their spatial position. The reason for this is that it is anticipated that the CFS-PML variables are spatially scaled, as specified in (6.37), (6.38), and (6.56). Therefore, the coefficients b and c are one-dimensional functions of these variables. Then, $b_{x_{i}}$ is the value of b_{x} (from (6.85)) computed at $x = i \Delta x$. Similarly, $b_{x}{}_{i+\frac{1}{2}}$ is the value of b_{x} computed at $x = (i + \frac{1}{2})\Delta x$.

6.5.3   EXAMPLE OF THE CFS-PML

The performance of the CFS-PML can be illustrated through a case study. Consider the patch antenna array fed by a 50-Ω microstrip line that is illustrated in Fig. 6.4. The scattering parameters

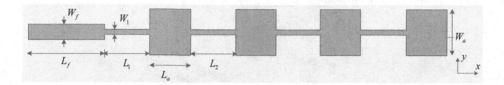

Figure 6.4: Microstrip fed antenna array printed on a conductor backed dielectric substrate with thickness $h = 1.5748$ mm and dielectric constant $\varepsilon_r = 2.1$. ($W_f = 3.93$ mm, $W_1 = 1.3$ mm, $W_a = 11.79$ mm, $L_f = 25.2$ mm, $L_1 = 13.1$ mm, $L_2 = 12.1$ mm, and $L_a = 10.08$ mm [1].)

of this array were computed via the FDTD method with a uniformly spaced rectangular grid. (Refer to Section 8.2.3 for more information on scattering parameters.) The spatial grid dimensions were $\Delta x = 1.008$ mm, $\Delta y = 0.3275$ mm, and $\Delta z = 0.2624667$ mm. The time-step was 0.99 times the Courant limit. Five of the exterior boundaries were terminated with a PML. The conducting ground plane backing the dielectric substrate was modeled as a PEC. Note that the substrate extended through the side PML layers to the terminating PEC boundary. This simulates that the substrate is effectively infinite in extent. The microstrip feed line was also extended through the PML to the terminating boundary. This simulates a matched line impedance.

Figure 6.5 illustrates the scattering parameters computed via the FDTD method based on three different simulations. The first was a reference simulation. In this simulation, the outer truncation boundary of the PML was extended out 50 space cells in all directions beyond the antenna. The boundary was then terminated with a 10-cell CFS-PML. The global solution space for this problem had a grid of dimension $219 \times 138 \times 57$ cells. This is considered the "reference" solution, in that the reflection error due to the PML is negligible.

The circuit was also simulated with a much smaller working volume using a 10-cell thick CFS-PML, and a 10-cell thick non-CFS CPML. In both cases grid dimensions remained exactly the same. However, the working volume extended only 1 cell beyond the edges of the antenna on the sides, and only 1 cell above the antenna. A 10 cell PML was then extruded beyond these points. The global dimension of the problem is $139 \times 58 \times 17$ spatial cells. For both the CFS-PML and the non-CFS-PML case, σ and κ were polynomial scaled with $m = 3$. In all cases, σ^{opt} was chosen using (6.43). For the side-walls, the dielectric constant was an average value of the free-space and substrate. For the ceiling, free-space was assumed. For the side wall with the microstrip feed, κ^{max} was set to 1 and a to 0. The reason for this is that this wall is predominately absorbing the TEM waves supported by the microstrip line. The other side walls had $\kappa^{max} = 15$, and $a^{max} = 0.2$. a was polynomial scaled via (6.9) with $m = 1$.

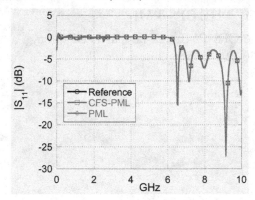

Figure 6.5: Scattering parameters of the antenna array in Fig. 6.4 computed via the FDD method with a reference solution, and with CFS-PML and PML ($a = 0$) placed only 1 cell from the antenna surface.

The excitation for the structure was a soft voltage source, with a Gaussian pulse waveform with a 10 GHz bandwidth, and delay $t_o = 4t_w$.

The error in the scattering parameters relative to the reference solution is illustrated in Fig. 6.6. The CFS-PML, on average, has errors that are 2 orders of magnitude smaller than the non-CFS-PML. At higher frequencies, there is still an order of magnitude improvement.

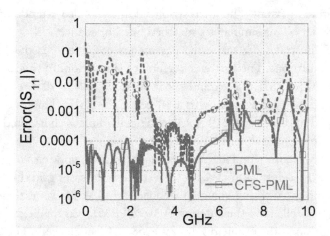

Figure 6.6: Relative error in the scattering parameters computed via the FDTD with the traditional PML ($a = 0$), and the CFS-PML.

Given that the PML was placed only 1 cell away from the antenna provides testimony of the effectiveness of this method as an absorber. The CFS-PML formulation enhances the effectiveness of the aborbing layer for absorbing evanescent waves present in the near field, and for providing a more transparent interface for low-frequency waves. Even though the PML formulation adds auxiliary variables, this is far offset by enabling the outer truncation boundary to be placed so close to a device under test.

6.6 PROBLEMS

1. Derive the wave equation (6.7) from (6.5) for the uniaxial anisotropic medium. Also, derive (6.8) from this equation and, consequently, the dispersion relations (6.9)–(6.10).

2. For the uniaxial anisotropic PML formulation in Section 6.2, solve the half-space problem for a TM_z-polarized wave. Show that the TM_z-polarized case also leads to a reflectionless interface if $\varepsilon_2 = \varepsilon_1, \mu_2 = \mu_1, b = d$ and $c^{-1} = d$.

3. Starting from Maxwell's curl equations in the stretched coordinate space, (6.28) and (6.29), for an x-normal PML media (e.g., $s_y = s_z = 1$), derive the wave equation for H_z for the TE_z-polarized case. Derive the dispersion relation (6.32). Also, show that (6.30) is the solution for H_z. Finally, derive (6.33).

4. Assume that s_x is only a function of x, and similarly assume $s_y(y)$ and $s_z(z)$. Then, by substituting $E_x = \tilde{E}_x/s_x, E_y = \tilde{E}_y/s_y, E_z = \tilde{E}_z/s_z, H_x = \tilde{H}_x/s_x, H_y = \tilde{H}_y/s_y$, and $H_z = s_z\tilde{H}_z$ into (6.28) and (6.29), show that this leads exactly to the anisotropic PML expression (6.5) with the tensor $\bar{\bar{s}}$ defined by (6.25).

5. Derive (6.59) and (6.60).

6. Derive (6.75) and (6.78).

7. Derive (6.84) and (6.85).

8. Write a computer program that models Maxwell's equations with a CFS-PML based on Equations (6.75)–(6.97) (c.f., Appendix B). To validate the code, try to reproduce the results in Fig. 6.5.

REFERENCES

[1] C. F. Wang, J. M. Jin, L. W. Li, P. S. Kooi, and M. S. Leong, "A multilevel BCG-FFT method for the analysis of a microstrip antenna and array," *Microwave and Optical Technology Letters*, vol. 27, pp. 20–23, Oct 5 2000.
DOI: 10.1002/1098-2760(20001005)27:1%3C20::AID-MOP7%3E3.0.CO;2-A 131

[2] J.-P. Berenger, "A perfectly matched layer for the absorption of electromagnetic waves," *Journal of Computational Physics*, vol. 114, pp. 195–200, 1994. DOI: 10.1006/jcph.1994.1159 113, 117

[3] J. P. Berenger, *Perfect Matched Layer (PML) for Computational Electromagnetics*: Morgan and Claypool Publishers, 2007. DOI: 10.2200/S00030ED1V01Y200605CEM008 113, 117, 125

[4] W. C. Chew and W. H. Weedon, "A 3D Perfectly Matched Medium from Modified Maxwells Equations with Stretched Coordinates," *Microwave and Optical Technology Letters*, vol. 7, pp. 599–604, 1994. DOI: 10.1002/mop.4650071304 113, 118

[5] Z. S. Sacks, D. M. Kingsland, R. Lee, and J. F. Lee, "A perfectly matched anisotropic absorber for use as an absorbing boundary condition," *IEEE Transactions on Antennas and Propagation*, vol. 43, pp. 1460–1463, 1995. DOI: 10.1109/8.477075 113

[6] S. D. Gedney, "An anisotropic perfectly matched layer-absorbing medium for the truncation of FDTD lattices," *IEEE Transactions on Antennas and Propagation*, vol. 44, pp. 1630–1639, December 1996. DOI: 10.1109/8.546249 113, 122

[7] J. A. Roden and S. D. Gedney, "Convolutional PML (CPML): An Efficient FDTD Implementation of the CFS-PML for Arbitrary Media," *Microwave and Optical Technology Letters*, vol. 27, pp. 334–339, December 5 2000.
DOI: 10.1002/1098-2760(20001205)27:5%3C334::AID-MOP14%3E3.3.CO;2-1 113, 124, 125, 126, 129

[8] C. A. Balanis, *Advanced Engineering Electromagnetics*. New York: Wiley, 1989. 115

[9] S. D. Gedney, "The Perfectly Matched Layer Absorbing Medium," in *Advances in Computational Electrodynamics: The Finite Difference Time Domain*, A. Taflove, Ed. Boston: Artech House, 1998, pp. 263–340. 122

[10] M. Kuzuoglu and R. Mittra, "Frequency dependence of the constitutive parameters of causal perfectly matched anisotropic absorbers," *IEEE Microwave and Guided Wave Letters*, vol. 6, pp. 447–449, 1996. DOI: 10.1109/75.544545 124

[11] S. D. Gedney, "Scaled CFS-PML: It is more robust, more accurate, more efficient, and simple to implement. Why aren't you using it?," in *2005 IEEE Antennas and Propagation Society International Symposium*, Washington, DC, 2005, pp. 364 - 367. DOI: 10.1109/APS.2005.1552824 125

[12] J. P. Berenger, "Numerical reflection from FDTD-PMLs: A comparison of the split PML with the unsplit and CFSPMLs," *IEEE Transactions on Antennas and Propagation*, vol. 50, pp. 258–265, Mar 2002. DOI: 10.1109/8.999615 125

[13] J. P. Berenger, "Application of the CFSPML to the absorption of evanescent waves in waveguides," *IEEE Microwave and Components Letters*, vol. 12, pp. 218–220, June 2002. DOI: 10.1109/LMWC.2002.1010000 125

[14] E. Becache, P. G. Petropoulos, and S. D. Gedney, "On the long-time behavior of unsplit perfectly matched layers," *IEEE Transactions On Antennas And Propagation,* vol. 52, pp. 1335–1342, May 2004. DOI: 10.1109/TAP.2004.827253 125

[15] S. D. Gedney and B. Zhao, "An Auxiliary Differential Equation Formulation for the Complex-Frequency Shifted PML," *IEEE Transactions on Antennas and Propagation,* vol. 58, 2010. DOI: 10.1109/TAP.2009.2037765 125

CHAPTER 7

Subcell Modeling

7.1 INTRODUCTION

The FDTD method is a powerful method for simulating the electromagnetic properties of practical problems. However, FDTD methods based on uniformly spaced orthogonal grids are challenged when attempting to simulate large complex structures that have magnitudes of scale that separate coarse geometric features from fine features. For example, consider the model of an entire aircraft that includes fine details such as seams in the fuselage, wire antennas, wire harnesses, and printed circuit boards in the on-board electronics. A high fidelity model would require a very finely spaced grid. Due to the uniformity of the gridding, small grid cells would also be forced to represent the coarse detail, such as the fuselage. This dramatically increases the grid size. Furthermore, due to the Courant limit, the time-step would be unreasonably small.

A class of methods referred to as "subcell" models have been introduced to improve the local grid fidelity of fine geometric detail without the constraint of a globally refined grid. Subcell models attempt to accurately model the electromagnetic fields near local geometrical features that are finer than the grid cell. This is accomplished with local approximations of the fields and often a local deformation of the rectangular grid. For example, a thin-wire antenna can be modeled via a thin-wire subcell model such that the antenna can be accurately assessed without having to reduce the dimension of the global grid to resolve the curvature of the thin wire.

In this chapter, a number of subcell models are introduced, starting with the thin-wire model. Narrow slots and thin material sheets are also discussed. The lumped circuit element models in Section 4.4.3 can also be considered as subcell models. Techniques to accurately resolve geometries (conducting and non-conducting) that do not conform to a rectangular grid based on the conformal patch modeling method are also introduced.

7.2 THIN WIRES

7.2.1 THE BASIC THIN-WIRE SUBCELL MODEL

A thin wire is defined as a wire with a radius, a, that is much smaller than its length, and is also much smaller than a wavelength. The thin-wire subcell model also assumes that the wire radius is small compared to the grid cell dimension (the maximum limit is $a < \Delta/2$). For simplicity, the axis of the thin wire is assumed to lie on primary grid edges of the rectangular lattice. For example, Fig. 7.1 illustrates a section of a thin wire, where the wire axis is coincident with a z-directed primary grid edge. Also illustrated is the primary grid face with a y-normal that shares the edge. The wire

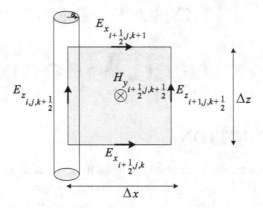

Figure 7.1: Thin wire with axis lying on a z-directed primary grid edge and the local primary grid face.

is assumed to be perfectly conducting, thus the tangential electric field is zero on the wire surface. In this example, $E_z = 0$ on the wire surface. The subcell model assumes that the local primary grid face is deformed to have a reduced width of $(\Delta x - a)$ to exclude the region inside the thin wire.

The electric current induced on the thin wire will radiate a magnetic field. Based on Ampère's law, it is known that the local magnetic field will have a $1/r$ type of singular behavior [1]. The electric field normal to the wire surface will also have a $1/r$ behavior. Thus, to more accurately represent the local fields, this singular behavior is applied to the local fields. Referring to Fig. 7.1, the tangential magnetic and the normal electric fields are expressed as [2]:

$$H_{y\,i+\frac{1}{2},j,k+\frac{1}{2}}^{n} = \tilde{H}_{y\,i+\frac{1}{2},j,k+\frac{1}{2}}^{n} \frac{\Delta x}{2} \frac{1}{|x - i\,\Delta x|} \tag{7.1}$$

and

$$E_{x\,i+\frac{1}{2},j,k}^{n+\frac{1}{2}} = \tilde{E}_{x\,i+\frac{1}{2},j,k}^{n+\frac{1}{2}} \frac{\Delta x}{2} \frac{1}{|x - i\,\Delta x|} \tag{7.2}$$

where $\tilde{H}_{y\,i+\frac{1}{2},j,k+\frac{1}{2}}^{n}$ and $\tilde{E}_{x\,i+\frac{1}{2},j,k}^{n+\frac{1}{2}}$ are the local unknown coefficients representing the fields at their respective discrete coordinate. Faraday's law in the integral form is then enforced over the local primary grid face:

$$-\mu \iint_S \frac{\partial \vec{H}}{\partial t} \cdot d\vec{s} = \oint_c \vec{E} \cdot d\vec{\ell}. \tag{7.3}$$

The singular field approximations of (7.1) and (7.2) are applied. $E_z\Big|_{i+1,j,k+\frac{1}{2}}$ is assumed to be constant along the vertical edge. Integrating over the discrete face cell, (7.3) can be expressed as:

$$-\mu\left[\tilde{H}_y^{n+1}\Big|_{i+\frac{1}{2},j,k+\frac{1}{2}} - \tilde{H}_y^n\Big|_{i+\frac{1}{2},j,k+\frac{1}{2}}\right]\frac{\Delta z}{\Delta t}\frac{\Delta x}{\Delta 2}\ln\left(\frac{\Delta x}{a}\right) =$$
$$-E_z^{n+\frac{1}{2}}\Big|_{i+1,j,k+\frac{1}{2}}\Delta z + \left[\tilde{E}_x^{n+\frac{1}{2}}\Big|_{i+\frac{1}{2},j,k+1} - \tilde{E}_x^{n+\frac{1}{2}}\Big|_{i+\frac{1}{2},j,k}\right]\frac{\Delta x}{2}\ln\left(\frac{\Delta x}{a}\right) \qquad (7.4)$$

where a central difference approximation of the time-derivative was performed, and $E_z\Big|_{i,j,k+\frac{1}{2}}$ is assumed to be zero because of the boundary condition on the wire surface. This leads to the explicit time-update equation:

$$\tilde{H}_y^{n+1}\Big|_{i+\frac{1}{2},j,k+\frac{1}{2}} = \tilde{H}_y^n\Big|_{i+\frac{1}{2},j,k+\frac{1}{2}} + \frac{\Delta t}{\mu}\left[\frac{E_z^{n+\frac{1}{2}}\Big|_{i+1,j,k+\frac{1}{2}}}{\frac{\Delta x}{2}\ln\left(\frac{\Delta x}{a}\right)} - \left(\frac{\tilde{E}_x^{n+\frac{1}{2}}\Big|_{i+\frac{1}{2},j,k+1} - \tilde{E}_x^{n+\frac{1}{2}}\Big|_{i+\frac{1}{2},j,k}}{\Delta z}\right)\right] \qquad (7.5)$$

which can be used to update the local magnetic field. On the opposite side of the wire:

$$\tilde{H}_y^{n+1}\Big|_{i-\frac{1}{2},j,k+\frac{1}{2}} = \tilde{H}_y^n\Big|_{i-\frac{1}{2},j,k+\frac{1}{2}} + \frac{\Delta t}{\mu}\left[\frac{-E_z^{n+\frac{1}{2}}\Big|_{i-1,j,k+\frac{1}{2}}}{\frac{\Delta x}{2}\ln\left(\frac{\Delta x}{a}\right)} - \left(\frac{\tilde{E}_x^{n+\frac{1}{2}}\Big|_{i-\frac{1}{2},j,k+1} - \tilde{E}_x^{n+\frac{1}{2}}\Big|_{i-\frac{1}{2},j,k}}{\Delta z}\right)\right]. \qquad (7.6)$$

The same analogy can be used for the H_x-updates. For example,

$$\tilde{H}_x^{n+1}\Big|_{i,j+\frac{1}{2},k+\frac{1}{2}} = \tilde{H}_x^n\Big|_{i,j+\frac{1}{2},k+\frac{1}{2}} + \frac{\Delta t}{\mu}\left[\left(\frac{\tilde{E}_y^{n+\frac{1}{2}}\Big|_{i,j+\frac{1}{2},k+1} - \tilde{E}_y^{n+\frac{1}{2}}\Big|_{i,j+\frac{1}{2},k}}{\Delta z}\right) - \frac{E_z^{n+\frac{1}{2}}\Big|_{i,j+1,k+\frac{1}{2}}}{\frac{\Delta y}{2}\ln\left(\frac{\Delta y}{a}\right)}\right], \qquad (7.7)$$

and similarly for $\tilde{H}_x^{n+\frac{1}{2}}\Big|_{i,j-\frac{1}{2},k+\frac{1}{2}}$.

7.2.2 CURVATURE CORRECTION

The local updates of the tangential magnetic fields using the basic thin-wire subcell model as expressed in (7.5)–(7.7) provide a good approximation for the thin-wire problem. It was shown by Mäkinen, et al., [3] that this model can be improved by accounting for the curvature of radius of

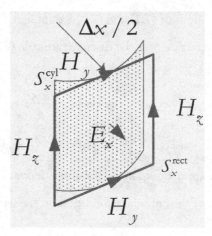

Figure 7.2: Primary grid face adjacent to the thin-wire surface with $\hat{x}$ normal and the effective cylindrical face (shaded) with constant radius of curvature that intersects the primary face center.

integration. Consider the primary grid face with x-normal as illustrated in Fig. 7.2. Ampere's law in its integral form is enforced over the face:

$$\frac{\partial}{\partial t}\varepsilon \iint_{s} \vec{E} \cdot d\vec{S} = \oint_{C} \vec{H} \cdot d\vec{\ell}. \tag{7.8}$$

The Yee-algorithm assumes that the normal electric field is constant over the rectangular primary grid face. However, E_x and H_y are assumed to have $1/r$ dependences. As a consequence, E_x is not constant over the secondary face, and H_y is not constant over the horizontal edges. Rather, they are constant over the cylindrical surface, which is the shaded surface in Fig. 7.2. Thus, the surface integral is more accurately expressed as:

$$\iint_{S_x^{rect}} \hat{r} E_r \cdot \hat{x} ds \tag{7.9}$$

where $\hat{r}$ is the radial unit vector of the cylindrical coordinate system. The radial field is assumed to have a $1/r$ variation, which will be explicitly included in this integration. The field is assumed to be constant along the z-direction. Thus, this reduces to a one-dimensional integration. The integral is also symmetric about $y = 0$, and thus can be written as [3]:

$$2\Delta z \tilde{E}_x \int_{0}^{\frac{\Delta y}{2}} \cos\phi(y) \frac{\frac{\Delta x}{2}}{\sqrt{\left(\frac{\Delta x}{2}\right)^2 + y^2}} dy = \Delta z \Delta x \tilde{E}_x \tan^{-1}\left(\frac{\Delta y}{\Delta x}\right) \tag{7.10}$$

where $\tilde{E}_x$ was expressed via (7.2) and $\hat{r} \cdot \hat{x} = \cos\phi$, where ϕ is the angle between $\hat{r}$ and the x-axis. ϕ is a function of y.

We can similarly treat the line integral of the magnetic field along the y-directed contour as:

$$2\tilde{H}_y \int_0^{\frac{\Delta y}{2}} \cos\theta(y) \frac{\frac{\Delta x}{2}}{\sqrt{\left(\frac{\Delta x}{2}\right)^2 + y^2}} dy = \Delta x \tilde{H}_y \tan^{-1}\left(\frac{\Delta y}{\Delta x}\right) \tag{7.11}$$

where H_ϕ is expressed via (7.1) and $\hat{\phi} \cdot \hat{y} = \cos\theta$, where θ is the angle between $\hat{\phi}$ and the y-axis. To reduce the notation, the scaling factor is introduced

$$h_x = \frac{\Delta x}{\Delta y} \tan^{-1}\left(\frac{\Delta y}{\Delta x}\right). \tag{7.12}$$

Therefore, combining (7.10) and (7.11) with (7.8) leads to

$$\varepsilon \Delta z \Delta y h_x \left(\frac{\tilde{E}_x^{n+\frac{1}{2}}{}_{i+\frac{1}{2},j,k} - \tilde{E}_x^{n-\frac{1}{2}}{}_{i+\frac{1}{2},j,k}}{\Delta t} \right) =$$

$$\Delta y h_x \left(\tilde{H}_y^n{}_{i+\frac{1}{2},j,k-\frac{1}{2}} - \tilde{H}_y^n{}_{i+\frac{1}{2},j,k+\frac{1}{2}} \right) + \Delta z \left(H_z^n{}_{i+\frac{1}{2},j+\frac{1}{2},k} - H_z^n{}_{i+\frac{1}{2},j-\frac{1}{2},k} \right) \tag{7.13}$$

in the discrete space. The electric field is again normalized as

$$\bar{E}_x = h_x \tilde{E}_x. \tag{7.14}$$

Consequently, (7.13) leads to the update equation:

$$\bar{E}_x^{n+\frac{1}{2}}{}_{i+\frac{1}{2},j,k} = \bar{E}_x^{n-\frac{1}{2}}{}_{i+\frac{1}{2},j,k} + \frac{\Delta t}{\varepsilon \Delta y}\left(H_z^n{}_{i+\frac{1}{2},j+\frac{1}{2},k} - H_z^n{}_{i+\frac{1}{2},j-\frac{1}{2},k} \right)$$

$$- \frac{\Delta t}{\varepsilon \Delta z} h_x \left(\tilde{H}_y^n{}_{i+\frac{1}{2},j,k+\frac{1}{2}} - \tilde{H}_y^n{}_{i+\frac{1}{2},j,k-\frac{1}{2}} \right). \tag{7.15}$$

Since $\bar{E}_x$ is now the unknown, the update for $\tilde{H}_y$ in (7.5) must also be modified. This is expressed as:

$$\tilde{H}_y^{n+1}{}_{i+\frac{1}{2},j,k+\frac{1}{2}} = \tilde{H}_y^n{}_{i+\frac{1}{2},j,k+\frac{1}{2}} + \frac{\Delta t}{\mu}\left[\frac{E_z^{n+\frac{1}{2}}{}_{i+1,j,k+\frac{1}{2}}}{\frac{\Delta x}{2}\ln\left(\frac{\Delta x}{a}\right)} - \left(\frac{\bar{E}_x^{n+\frac{1}{2}}{}_{i+\frac{1}{2},j,k+1} - \bar{E}_x^{n+\frac{1}{2}}{}_{i+\frac{1}{2},j,k}}{h_x \Delta z} \right) \right]. \tag{7.16}$$

Following a similar analogy for the y-directed surface, the y-directed electric field near the thin wire is also normalized again as:

$$\bar{E}_y = h_y \tilde{E}_y \tag{7.17}$$

where

$$h_y = \frac{\Delta y}{\Delta x} \tan^{-1} \left(\frac{\Delta x}{\Delta y} \right). \tag{7.18}$$

The update for the y-directed field then becomes:

$$\bar{E}_{y\,i,j+\frac{1}{2},k}^{n+\frac{1}{2}} = \bar{E}_{y\,i,j+\frac{1}{2},k}^{n-\frac{1}{2}} + \frac{\Delta t}{\varepsilon \Delta z} h_y \left(\tilde{H}_{x\,i,j+\frac{1}{2},k+\frac{1}{2}}^{n} - \tilde{H}_{x\,i,j+\frac{1}{2},k-\frac{1}{2}}^{n} \right)$$
$$- \frac{\Delta t}{\varepsilon \Delta x} \left(H_{z\,i+\frac{1}{2},j+\frac{1}{2},k}^{n} - H_{z\,i-\frac{1}{2},j+\frac{1}{2},k}^{n} \right). \tag{7.19}$$

The update for H_x in (7.7) must then be modified to account for the electric field normalization, leading to:

$$\tilde{H}_{x\,i,j+\frac{1}{2},k+\frac{1}{2}}^{n+1} = \tilde{H}_{x\,i,j+\frac{1}{2},k+\frac{1}{2}}^{n} + \frac{\Delta t}{\mu} \left[\left(\frac{\bar{E}_{y\,i,j+\frac{1}{2},k+1}^{n+\frac{1}{2}} - \bar{E}_{y\,i,j+\frac{1}{2},k}^{n+\frac{1}{2}}}{h_y \Delta z} \right) - \frac{E_{z\,i,j+1,k+\frac{1}{2}}^{n+\frac{1}{2}}}{\frac{\Delta y}{2} \ln \left(\frac{\Delta y}{a} \right)} \right]. \tag{7.20}$$

The modified update scheme expressed in (7.15)–(7.20) will provide a more accurate approximation for the thin-wire subcell model.

7.2.3 MODELING THE END-CAP

When modeling open-ended thin-wires, some care must be taken near the end of the wire. At the end of the wire, the currents don't simply go to zero. Rather, the currents flow on the end-cap in a radial direction and cancel to zero at the origin. The end-cap current can have a significant impact on the local electromagnetic fields. This is especially true for a wire with a thicker radius. An effective way of modeling the end-cap current is to introduce a line-charge at the top of the thin wire. The line charge must satisfy the conservation of charge:

$$\iint_S \vec{J} \cdot d\vec{s} = -\frac{d}{dt} \oint_c \rho \, dl \tag{7.21}$$

where ρ is the line charge density, and $\vec{J}$ is the current density on the wire surface. The line charge density is assumed to be constant around the contour bounding the end-cap. Then, (7.21) can be approximated as:

$$I = -\frac{\Delta q}{\Delta t} \tag{7.22}$$

where I is the total current flowing from the wire into the end cap, and Δq is the change in the total charge at the end-cap rim per unit time Δt. The current is computed via a quasi-static approximation:

$$I = \oint_C \vec{H} \cdot d\vec{\ell}. \tag{7.23}$$

The integral is performed over a rectangular contour that is the secondary-grid face, centered about the wire axis and $1/2$ of a cell below the end-cap. Again, for simplicity, the wire is assumed to be aligned with the z-axis. Thus,

$$I^n_{k_{top}-\frac{1}{2}} = \Delta y h_x \left(\tilde{H}^n_{y,i+\frac{1}{2},j,k_{top}-\frac{1}{2}} - \tilde{H}^n_{y,i-\frac{1}{2},j,k_{top}-\frac{1}{2}} \right)$$
$$- \Delta x h_y \left(\tilde{H}^n_{x,i,j+\frac{1}{2},k_{top}-\frac{1}{2}} - \tilde{H}^n_{x,i,j-\frac{1}{2},k_{top}-\frac{1}{2}} \right). \tag{7.24}$$

Then, from (7.22)

$$\frac{\Delta q^n_{top}}{\Delta t} \approx - \left[\Delta y h_x \left(\tilde{H}^n_{y,i+\frac{1}{2},j,k_{top}-\frac{1}{2}} - \tilde{H}^n_{y,i-\frac{1}{2},j,k_{top}-\frac{1}{2}} \right) \right.$$
$$\left. - \Delta x h_y \left(\tilde{H}^n_{x,i,j+\frac{1}{2},k_{top}-\frac{1}{2}} - \tilde{H}^n_{x,i,j-\frac{1}{2},k_{top}-\frac{1}{2}} \right) \right]. \tag{7.25}$$

Note that there is some error introduced by the fact that the current is measured a half-cell from the end-cap. However, in practice, this is found to be a good approximation.

The electrostatic charge ring will produce a scalar potential [4]:

$$\phi = \frac{1}{4\pi\varepsilon} \int_C \frac{\rho}{R} dl \tag{7.26}$$

where R is the distance from a point on the line charge to the observation coordinate. The electric field due to the quasi-static charge is then computed as:

$$E = -\nabla\phi. \tag{7.27}$$

The contour of integration is the circular ring defining the end-cap, as illustrated in Fig. 7.3. For this contour, the electric-scalar potential computed via (7.26) can be expressed in the local cylindrical coordinates with origin at the center of the end-cap as

$$\phi(r,z) = \frac{1}{4\pi\varepsilon} \frac{q}{2\pi a} \int_0^{2\pi} \frac{1}{\sqrt{r^2 + a^2 - 2ra\cos\phi + z^2}} a \, d\phi, \tag{7.28}$$

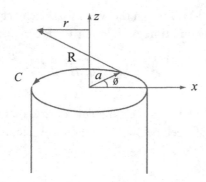

Figure 7.3: Contour C bounding the thin wire end-cap that is supporting the charge density ring.

where q is the total charge, which is assumed to be uniformly distributed over the ring. This can be computed analytically as:

$$\phi(r, z) = \frac{1}{4\pi\varepsilon} \frac{q}{2\pi a} \frac{4aK(k)}{\sqrt{(r+a)^2 + z^2}} \tag{7.29}$$

where K is a complete elliptic integral of the first kind, and the argument

$$k = \frac{2\sqrt{ar}}{\sqrt{(r+a)^2 + z^2}}. \tag{7.30}$$

The electric field due to the charge ring is then computed via (7.27). Analytical expressions for the axial and radial fields as derived by Mäkinen, et al., [3] are:

$$E_z^q(r, z) = -\frac{\partial\phi(r, z)}{\partial z} = \frac{q}{4\pi^2\varepsilon} \frac{zE(k)}{2\left((r-a)^2 + z^2\right)\sqrt{(r+a)^2 + z^2}} \tag{7.31}$$

and

$$E_r^q(r, z) = -\frac{\partial\phi(r, z)}{\partial r} = \frac{q}{4\pi^2\varepsilon} \frac{\left((r-a)^2 + z^2\right)K(k) + \left(r^2 - a^2 - z^2\right)zE(k)}{r\left((r-a)^2 + z^2\right)\sqrt{(r+a)^2 + z^2}}, \tag{7.32}$$

where $E(k)$ is the complete elliptic integral of the second kind with argument k as defined by (7.30).

The quasi-static electric field due to the charge ring can be included in the field updates for secondary-grid faces adjacent to the wire end-cap. Here, Ampere's law is expressed as:

$$\frac{\partial}{\partial t}\varepsilon \iint\limits_S \vec{E} \cdot d\vec{s} = \frac{\partial}{\partial t}\varepsilon \iint\limits_S \vec{E}^q \cdot d\vec{s} + \oint\limits_C \vec{H} \cdot d\vec{\ell} \tag{7.33}$$

where the electric field on the left-hand-side represents the *total* electric field. The expression requires the time derivative of $\vec{E}^q$. Referring to (7.31) and (7.32), it is seen that the charge is the time-dependent quantity. Consequently, we approximate:

$$\frac{\partial q}{\partial t} \approx \frac{\Delta q}{\Delta t} \tag{7.34}$$

which is defined by (7.25).

Now, consider the axial field update just above the end-cap, as illustrated in Fig. 7.4. That is,

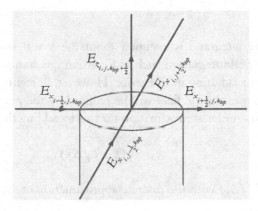

Figure 7.4: Near discrete-electric fields adjacent to the thin-wire end cap.

(7.33) is applied over the secondary grid face one-half of a cell above the top of the thin wire. This requires evaluating the normal projection of the quasi-static electric field over the cell surface, which is expressed as:

$$E_{z_{\text{int}}}^{q'} = 4 \int_0^{\Delta y/2} \int_0^{\Delta x/2} E_z^{q'}\left(\sqrt{x^2 + y^2}, \Delta z/2\right) dxdy \tag{7.35}$$

where $E_z^{q'}(r, z)$ is the time-derivative of $E_z^q(r, z)$ derived via (7.31) with q replaced by $\Delta q/\Delta t$. Note that the range of integration is reduced to one quadrant by taking advantage of the symmetry of the axial field. This integral can be evaluated via a numerical quadrature rule, such as Gaussian quadrature [5]. The numerical evaluation of the complete elliptic integrals can be found in Zhang and Jin [6]. Finally, (7.33) leads to the recursive update equation:

$$E_z^{n+\frac{1}{2}}_{i,j,k_{top}+\frac{1}{2}} = E_z^{n-\frac{1}{2}}_{i,j,k_{top}+\frac{1}{2}} + \frac{\Delta t}{\Delta x \Delta y} E_{z_{\text{int}}}^{q'n} +$$

$$\frac{\Delta t}{\varepsilon}\left[\frac{1}{\Delta x}\left(H_y^n_{i+\frac{1}{2},j,k_{top}+\frac{1}{2}} - H_y^n_{i-\frac{1}{2},j,k_{top}+\frac{1}{2}}\right) - \frac{1}{\Delta y}\left(H_x^n_{i,j+\frac{1}{2},k_{top}+\frac{1}{2}} - H_x^n_{i,j-\frac{1}{2},k_{top}+\frac{1}{2}}\right)\right]. \tag{7.36}$$

Note that $E_{z_{\text{int}}}^{q'\,n}$ is also a function of the magnetic field through (7.25).

Next, consider the radial fields. As an example, consider the secondary grid face with center $(i + \frac{1}{2}, j, k_{top})$. From (7.33), this requires the integral

$$E_{x_{\text{int}}}^{q'} = 2 \int\limits_{-\Delta z/2}^{\Delta z/2} \int\limits_{0}^{\Delta y/2} E_r^{q'} \left(\sqrt{(\Delta x/2)^2 + y^2}, z \right) \cos \phi \, dy \, dz \tag{7.37}$$

where symmetry in the integrand is assumed about the $y = 0$ axis. Again, this integral can be performed using numerical integration such as Gaussian quadrature [5]. One issue of note is that the ring is assumed to be radiating in free space. However, in reality, it is radiating in the presence of the thin wire. Consequently, the field will be reduced by the presence of the wire. Mäkinen, et al., [3] recommend a first-order approximation for this by adding the scaling factor:

$$k_M = (1 - a/\Delta y). \tag{7.38}$$

Combining (7.37) and (7.38) with the discrete approximation of (7.33) as derived by (7.13), leads to the update:

$$\bar{E}_x^{n+\frac{1}{2}}\bigg|_{i+\frac{1}{2},j,k} = \bar{E}_x^{n-\frac{1}{2}}\bigg|_{i+\frac{1}{2},j,k} + \frac{k_M \Delta t}{\Delta z \Delta y} E_{x_{\text{int}}}^{q'n} +$$

$$\frac{\Delta t}{\varepsilon} \left[\frac{1}{\Delta y} \left(H_z^n\bigg|_{i+\frac{1}{2},j+\frac{1}{2},k} - H_z^n\bigg|_{i+\frac{1}{2},j-\frac{1}{2},k} \right) - \frac{h_x}{\Delta z} \left(\tilde{H}_y^n\bigg|_{i+\frac{1}{2},j,k+\frac{1}{2}} - \tilde{H}_y^n\bigg|_{i+\frac{1}{2},j,k-\frac{1}{2}} \right) \right]. \tag{7.39}$$

Note that $E_{x_{\text{int}}}^{q'\,n}$ is a function of the magnetic field through (7.25).

Including the quasi-static field effects of the end-cap can significantly improve the thin wire approximation, and further reduce the overall discretization along the axis of the thin-wire.

7.2.4 DELTA-GAP SOURCE

When modeling a thin wire as an antenna, a common feed model is the delta-gap source – that is, a voltage applied across an infinitesimally narrow gap. Assume a z-directed wire axis. If the delta-gap source has a voltage of V-volts, the electric field in the gap can be expressed as [7]:

$$E_z|_{z_{gap}} = -V \delta(z - z_{gap}) \tag{7.40}$$

where δ is the "delta"-function. Referring to Fig. 7.5, the delta-gap source is assumed to be located on the edge $E_z\big|_{i,j,k+\frac{1}{2}}$. Consequently, Faraday's law (7.3) is expressed in discrete form as:

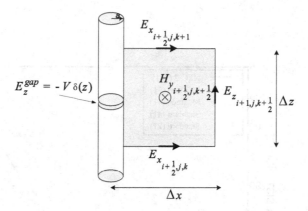

Figure 7.5: Thin wire with axis lying on a z-directed primary grid edge and the local primary grid face.

$$- \mu \left[\tilde{H}_y^{n+1}{}_{i+\frac{1}{2},j,k+\frac{1}{2}} - \tilde{H}_y^n{}_{i+\frac{1}{2},j,k+\frac{1}{2}} \right] \frac{\Delta z}{\Delta t} \frac{\Delta x}{2} \ln \left(\frac{\Delta x}{a} \right) =$$

$$\left[-V^{n+\frac{1}{2}} - E_z^{n+\frac{1}{2}}{}_{i+1,j,k+\frac{1}{2}} \Delta z \right] + \left[\tilde{E}_x^{n+\frac{1}{2}}{}_{i+\frac{1}{2},j,k+1} - \tilde{E}_x^{n+\frac{1}{2}}{}_{i+\frac{1}{2},j,k} \right] \frac{\Delta x}{2} \ln \left(\frac{\Delta x}{a} \right) \qquad (7.41)$$

where the term $-V^{n+\frac{1}{2}}$ is a consequence of the line integral over the surface of the thin wire. Also, substituting in the normalization (7.14), this leads to the explicit update equation:

$$\tilde{H}_y^{n+1}{}_{i+\frac{1}{2},j,k+\frac{1}{2}} = \tilde{H}_y^n{}_{i+\frac{1}{2},j,k+\frac{1}{2}} + \frac{\Delta t}{\mu} \left[\frac{E_z^{n+\frac{1}{2}}{}_{i+1,j,k+\frac{1}{2}} + \frac{V^{n+\frac{1}{2}}}{\Delta z}}{\frac{\Delta x}{2} \ln \left(\frac{\Delta x}{a} \right)} - \left(\frac{\tilde{E}_x^{n+\frac{1}{2}}{}_{i+\frac{1}{2},j,k+1} - \tilde{E}_x^{n+\frac{1}{2}}{}_{i+\frac{1}{2},j,k}}{h_x \Delta z} \right) \right] \qquad (7.42)$$

where $V^{n+\frac{1}{2}}$ is the voltage at discrete time $(n + \frac{1}{2})\Delta t$.

As an example, consider a straight half-wave dipole antenna. The length of the antenna is $L = 0.305$ m. The dipole is center fed with a delta-gap source. The dipole was modeled using the thin-wire model presented in this section combined with a uniformly spaced FDTD discretization. Two cell sizes are considered: $\Delta = L/21$ and $\Delta = L/41$. Namely, 21 and 41 cells were used along the antenna length. The exterior mesh was terminated via a PML absorbing layer.

The input impedance of the dipole at the frequency $f = c/2L$ (where c is the speed of light in free space) was computed with the thin-wire model based FDTD method for a range of wire radii [3]. The results are illustrated in Figs. 7.6 (a) and (b) as reported from [3]. Reference results

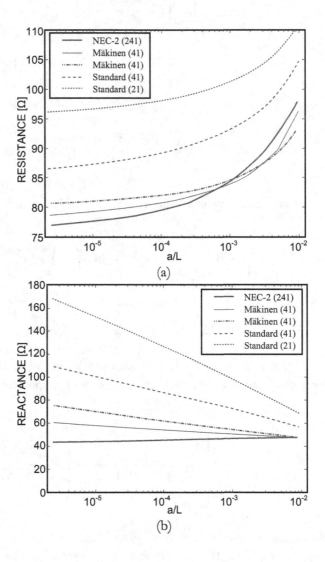

Figure 7.6: Input impedance of the dipole antenna at $f = c/2L$ (where $L = 0.305$ m) as a function of the dipole radius (expressed as a/L) computed using the "standard model" of Section 7.2.1 and Mäkinen's model, detailed in Section 7.2.3. (a) The input resistance. (b) The input reactance. *Source:* Mäkinen, et al., *IEEE Trans. Microwave Theory and Techniques*, 2002, pp. 1245–1255. © 2002 IEEE.

computed by the Numerical Electromagnetics Code (NEC) [8] are also included in these graphs. Two different models are illustrated – the "standard-model," which is based on the model presented in Section 7.2.1, and the Mäkinen's model, which is presented in Sections 7.2.2 and 7.2.3. Clearly, Mäkinen's model significantly improves the input impedance. The case with only 41 cells along the antenna's length is quite accurate, as compared with the NEC results, up to a radius of $a/L = 10^{-3}$.

Next, consider the effect of the end-cap on the thin-wire model. The dipole has a radius $a = 0.1\Delta$ and $\Delta = L/41$. Figure 7.7 illustrates the magnitude of the scattering parameter $|S_{11}|$ (magnitude of the reflection coefficient), referenced to 50 Ω, of the center-fed dipole antenna as a function of frequency. The results are again compared to NEC-2 results, Mäkinen's thin-wire model without the end-cap approximation (Section 7.2.3), and the "standard-model" of Section 7.2.1. Also, in this graph is Mäkinen's model without the end-cap approximation, which is based on the method of Section 7.2.2. It is apparent that accounting for the curvature does improve the results as compared to the "standard method." However, including the end-cap model leads to a dramatic improvement in the results. In fact, the full thin-wire model accurately predicts the resonance and the magnitude of the scattering parameter, whereas the other models still have significant error.

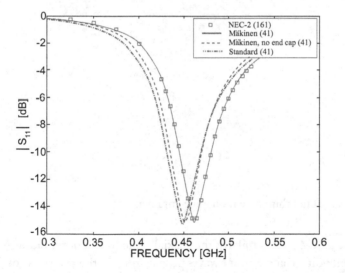

Figure 7.7: Effect of the end-cap model on $|S_{11}|$ of the 0.305 m dipole antenna with radius $a = 0.1\Delta$, where $\Delta = L/41$. $|S_{11}|$ is reference to 50 Ω. Results compute via NEC-2, Mäkinen's model, detailed in Section 7.2.3. and Mäkinen's model without the end-cap, detailed in Section 7.2.2, and the standard model of Section 7.2.1 are presented. *Source:* Mäkinen, et al., *IEEE Trans. Microwave Theory and Techniques,* 2002, pp. 1245–1255. © 2002 IEEE.

7.2.5   A TRANSMISSION LINE FEED

Often a wire antenna is excited by a transmission line feed. For example, a monopole antenna can be excited directly by a coaxial waveguide. In this case, typically, the excitation frequency is low enough such that only the fundamental TEM mode is excited. The propagation of this mode can be accurately modeled via a one-dimensional transmission line feed model. The one-dimensional transmission line model is then terminated by a connection to the three-dimensional FDTD algorithm at the feed location. This method developed by Maloney, et al., [9] will be outlined in this section.

As an illustration of the transmission line feed model, consider the coaxial-line feed excitation of a monopole antenna as illustrated in Fig. 7.8. The monopole is assumed to be a thin wire with

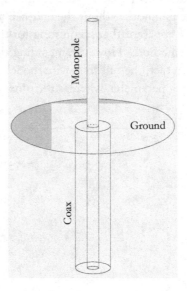

Figure 7.8: Coaxial fed thin-wire monopole antenna.

radius a, where $a < \Delta x/2$. Similarly, the coaxial cable has an outer radius b, where $b < \Delta x/2$. The time-dependent voltage and current waves propagating on the transmission line can be modeled via the one-dimensional FDTD method described in Chapter 2. To this end, the transmission line is also discretized with a uniformly spaced grid. The update equations presented in (2.13) and (2.14) derived from the discrete transmission line equations are expressed as:

$$V_i^{n+\frac{1}{2}} = V_i^{n-\frac{1}{2}} - Z_o \frac{v\Delta t}{\Delta z} \left(I_{i+\frac{1}{2}}^n - I_{i-\frac{1}{2}}^n \right) \tag{7.43}$$

$$I_{i+\frac{1}{2}}^{n+1} = I_{i+\frac{1}{2}}^n - \frac{1}{Z_o} \frac{v\Delta t}{\Delta z} \left(V_{i+1}^{n+\frac{1}{2}} - V_i^{n+\frac{1}{2}} \right) \tag{7.44}$$

where Z_o is the characteristic line impedance and v is the wave speed of the coaxial transmission line. These are defined in terms of the capacitance and inductance per unit length of the line as:

$$Z_o = \sqrt{\frac{L}{C}}, \quad v = \frac{1}{\sqrt{LC}}. \tag{7.45}$$

The incident wave can be launched onto the transmission line via a Thévenin voltage source, as presented in Section 2.6. By choosing the generator impedance $R_g = Z_o$, the transmission line is matched. The time-step Δt and the spatial cell size Δz of the transmission line model is assumed to be identical to that of the three-dimensional FDTD model. It is noted that this is within the stability limit of the one-dimensional FDTD model.

The one-dimensional transmission line is coupled to the three-dimensional FDTD model by attaching the terminating voltage node of the one-dimensional transmission line to a primary grid edge of the three-dimensional FDTD model. This is illustrated in Fig. 7.9. Let $i = N$ be the terminating voltage node of the transmission line, with discrete voltage $V_N^{n+\frac{1}{2}}$, as illustrated in Fig. 7.10. Consider the update of the primary grid face in Fig. 7.9. The discrete form of Faraday's law is expressed as:

$$- \mu \left[\tilde{H}_{y}^{n+1}{}_{i+\frac{1}{2},j,k+\frac{1}{2}} - \tilde{H}_{y}^{n}{}_{i+\frac{1}{2},j,k+\frac{1}{2}} \right] \frac{\Delta z}{\Delta t} \frac{\Delta x}{2} \ln\left(\frac{\Delta x}{a}\right) =$$
$$\tilde{E}_{x}^{n+\frac{1}{2}}{}_{i+\frac{1}{2},j,k+1} \frac{\Delta x}{2} \ln\left(\frac{\Delta x}{a}\right) - E_{z}^{n+\frac{1}{2}}{}_{i+1,j,k+\frac{1}{2}} \Delta z - V_N^{n+\frac{1}{2}} \tag{7.46}$$

where the line integral over the edge cutting across the aperture is equated to the transmission line voltage. This is rearranged to form the explicit update equation:

$$\tilde{H}_{y}^{n+1}{}_{i+\frac{1}{2},j,k+\frac{1}{2}} = \tilde{H}_{y}^{n}{}_{i+\frac{1}{2},j,k+\frac{1}{2}} - \frac{\Delta t}{\mu} \left(\frac{\tilde{E}_{x}^{n+\frac{1}{2}}{}_{i+\frac{1}{2},j,k+1}}{h_x \Delta z} - \frac{E_{z}^{n+\frac{1}{2}}{}_{i+1,j,k+\frac{1}{2}}}{\frac{\Delta x}{2} \ln\left(\frac{\Delta x}{a}\right)} \right) + \frac{\Delta t}{\mu \Delta z} \frac{V_N^{n+\frac{1}{2}}}{\frac{\Delta x}{2} \ln\left(\frac{\Delta x}{a}\right)} \tag{7.47}$$

where $\bar{E}_x$ is the normalized field as defined by (7.14) and h_x is defined by (7.12). Similar equations are derived for the other three primary grid cell faces that share the edge $E_{z}{}_{i,j,k+\frac{1}{2}}$.

The last step is to couple the 3D model back to the 1D model. This is done by computing the current induced on the thin-wire monopole and coupling this back into the 1D transmission line model as an independent current source, as illustrated in Fig. 7.8. The current is computed via a quasi-static approximation of Ampere's law:

$$\iint_S \vec{J} \cdot d\vec{s} = \oint_C \vec{H} \cdot d\vec{\ell}. \tag{7.48}$$

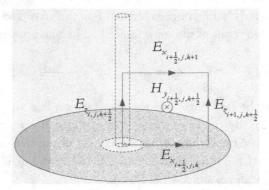

Figure 7.9: Primary cell face containing edges on the thin wire and cutting across the aperture of the coaxial transmission line.

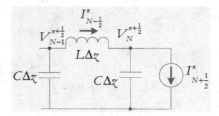

Figure 7.10: Terminating end of the one-dimensional coaxial transmission line model.

The surface integral is simply the total current I passing through S. The contour integral is evaluated with the aid of (7.11). Consequently, (7.48) is written in discrete form as:

$$I^n_{N+\frac{1}{2}} = \Delta y h_x \left(H^n_{y_{i+\frac{1}{2},j,k+\frac{1}{2}}} - H^n_{y_{i-\frac{1}{2},j,k+\frac{1}{2}}} \right) + \Delta x h_y \left(H^n_{x_{i,j-\frac{1}{2},k+\frac{1}{2}}} - H^n_{x_{i,j+\frac{1}{2},k+\frac{1}{2}}} \right). \qquad (7.49)$$

It is observed that the surface S, bound by the contour C, is actually displaced one half a cell (or distance $\Delta z/2$) above the coaxial aperture. This is actually consistent with the transmission-line model where the current is displaced by one-half of a space cell relative to the voltage.

7.3   CONFORMAL FDTD METHODS FOR CONDUCTING BOUNDARIES

The previous sections developed subcell models for specific geometries. However, there is a need to develop modeling techniques for more general geometries that do not conform to a rectangular grid. For example, consider the discrete representation of a sphere. The rectangular FDTD grid effectively represents the surface of the sphere with a "stair-case approximation." Improved resolution of the curved surface requires a significant reduction in the mesh size. Even halving the mesh dimensions increases the memory by a factor of 8, and commensurately the number of floating-point operations per time-step. The number of iterations will also double, since the time-step will be cut in half. Thus, overall, the computational time will increase by a factor of 16.

A more effective and efficient means to accurately model complex geometries is to employ a conformal FDTD method. Such a method is based on a local deformation of the grid to better approximate an arbitrary surface geometry. One of the first approaches to this method was proposed by Jurgens, et al. [10]. This method, referred to as the Contour-Path (CP)-FDTD method, did improve the accuracy. However, the method suffered from late-time instabilities. More recently, a number of approaches founded on the Dey-Mittra (DM) conformal FDTD algorithm have been proposed that overcome late-time stabilities, and provide the accuracy of a conformal grid [11]– [17]. Three of these methods are detailed in the following sections: The original Dey-Mittra (DM) algorithm [11, 12], the Yu-Mittra (YM) method [13, 17], which relaxes stability at the cost of accuracy, and the Benkler, Chavannes, and Kuster (BCK) method [15], which provides a balance between stability and accuracy.

7.3.1   DEY-MITTRA (DM) CONFORMAL FDTD METHOD FOR CONDUCTING BOUNDARIES

Consider a closed PEC object embedded within an orthogonal FDTD grid, as illustrated in Fig. 7.11. The shaded region represents the interior of PEC object where the electric and magnetic fields are

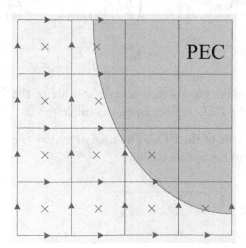

Figure 7.11: Cross section of a close three-dimensional PEC object embedded within a rectangular grid. The primary grid faces are shown.

zero. The primary mesh is deformed near the surface of the PEC. This impacts the magnetic-field update based on Faraday's law. This local deformation of a primary grid face is illustrated by Fig. 7.12. In this example, the PEC surface cuts across two of the primary grid edges. The line integral of the

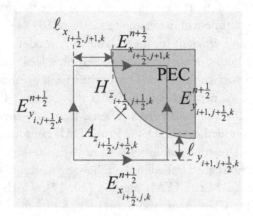

Figure 7.12: Local deformation of a primary grid face partially embedded in a closed PEC object. ℓ_x and ℓ_y are partial lengths and A_z is the partial area outside of the PEC region.

electric field over the edges cut by the PEC are non-zero outside of the PEC. The integral along the surface of the PEC is zero. Referring to Fig. 7.12, Faraday's law is approximated as:

$$\frac{\partial}{\partial t}\mu H_{z\,i+\frac{1}{2},j+\frac{1}{2},k}^{n+\frac{1}{2}} A_{z\,i+\frac{1}{2},j+\frac{1}{2},k} = -\left(E_{x\,i+\frac{1}{2},j,k}^{n+\frac{1}{2}} \ell_{x\,i+\frac{1}{2},j,k} + E_{y\,i+1,j+\frac{1}{2},k}^{n+\frac{1}{2}} \ell_{y\,i+1,j+\frac{1}{2},k} \right.$$

$$\left. - E_{x\,i+\frac{1}{2},j+1,k}^{n+\frac{1}{2}} \ell_{x\,i+\frac{1}{2},j+1,k} - E_{y\,i,j+\frac{1}{2},k}^{n+\frac{1}{2}} \ell_{y\,i,j+\frac{1}{2},k} \right) \tag{7.50}$$

where the ℓ_x and ℓ_y are the partial edge lengths outside of the PEC, and A_z is the area of the face that is outside of the PEC region. In this example $\ell_{x\,i+\frac{1}{2},j,k} = \Delta x$ and $\ell_{y\,i,j+\frac{1}{2},k} = \Delta y$, since these edges are fully outside of the PEC. The time derivative is evaluated with a central difference approximation, leading to the update equation:

$$H_{z\,i+\frac{1}{2},j+\frac{1}{2},k}^{n+1} = H_{z\,i+\frac{1}{2},j+\frac{1}{2},k}^{n} - \frac{\Delta t}{\mu A_{z\,i+\frac{1}{2},j+\frac{1}{2},k}} \cdot$$

$$\left(E_{y\,i+1,j+\frac{1}{2},k}^{n+\frac{1}{2}} \ell_{y\,i+1,j+\frac{1}{2},k} - E_{y\,i,j+\frac{1}{2},k}^{n+\frac{1}{2}} \ell_{y\,i,j+\frac{1}{2},k} - E_{x\,i+\frac{1}{2},j+1,k}^{n+\frac{1}{2}} \ell_{x\,i+\frac{1}{2},j+1,k} + E_{x\,i+\frac{1}{2},j,k}^{n+\frac{1}{2}} \ell_{x\,i+\frac{1}{2},j,k} \right). \tag{7.51}$$

Similar expressions can be derived for H_x and H_y.

The discrete electric fields are non-zero if their partial lengths $\ell > 0$. The electric fields are updated using the standard updates of (3.28)–(3.30). It is noted that any edge with partial length $\ell > 0$ will be shared by four faces with non-zero areas.

The update equation in (7.51) can be manipulated into a more convenient form. To this end, the second term on the right-hand-side of (7.51) is multiplied by the ratio $\Delta x \Delta y / \Delta x \Delta y$. Then, (7.51) is rewritten as:

$$
H_z^{n+1}\Big|_{i+\frac{1}{2},j+\frac{1}{2},k} = H_z^n\Big|_{i+\frac{1}{2},j+\frac{1}{2},k} - \frac{\Delta t}{\mu S_z\big|_{i+\frac{1}{2},j+\frac{1}{2},k}} \cdot
$$

$$
\left(\frac{E_y^{n+\frac{1}{2}}\big|_{i+1,j+\frac{1}{2},k}\, \delta_y\big|_{i+1,j+\frac{1}{2},k} - E_y^{n+\frac{1}{2}}\big|_{i,j+\frac{1}{2},k}\, \delta_y\big|_{i,j+\frac{1}{2},k}}{\Delta x} - \frac{E_x^{n+\frac{1}{2}}\big|_{i+\frac{1}{2},j+1,k}\, \delta_x\big|_{i+\frac{1}{2},j+1,k} - E_x^{n+\frac{1}{2}}\big|_{i+\frac{1}{2},j,k}\, \delta_x\big|_{i+\frac{1}{2},j,k}}{\Delta y} \right)
$$

$$(7.52)$$

where δ_x and δ_y are the fractional lengths, defined as:

$$
\delta_x\big|_{i+\frac{1}{2},j,k} = \frac{\ell_x\big|_{i+\frac{1}{2},j,k}}{\Delta x}, \quad \delta_x\big|_{i+\frac{1}{2},j+1,k} = \frac{\ell_x\big|_{i+\frac{1}{2},j+1,k}}{\Delta x},
$$

$$
\delta_y\big|_{i,j+\frac{1}{2},k} = \frac{\ell_y\big|_{i,j+\frac{1}{2},k}}{\Delta y}, \quad \delta_y\big|_{i+1,j+\frac{1}{2},k} = \frac{\ell_y\big|_{i+1,j+\frac{1}{2},k}}{\Delta y}, \tag{7.53}
$$

and S_z represents the fractional area:

$$
S_z\big|_{i+\frac{1}{2},j+\frac{1}{2},k} = \frac{A_z\big|_{i+\frac{1}{2},j+\frac{1}{2},k}}{\Delta x \Delta y}. \tag{7.54}
$$

Next, the electric fields are scaled by the fractional lengths. Namely, let:

$$
\tilde{E}_x^{n+\frac{1}{2}}\Big|_{i+\frac{1}{2},j,k} = E_x^{n+\frac{1}{2}}\Big|_{i+\frac{1}{2},j,k}\, \delta_x\big|_{i+\frac{1}{2},j,k},
$$

$$
\tilde{E}_y^{n+\frac{1}{2}}\Big|_{i,j+\frac{1}{2},k} = E_y^{n+\frac{1}{2}}\Big|_{i,j+\frac{1}{2},k}\, \delta_y\big|_{i,j+\frac{1}{2},k}, \quad \tilde{E}_z^{n+\frac{1}{2}}\Big|_{i,j,k+\frac{1}{2}} = E_z^{n+\frac{1}{2}}\Big|_{i,j,k+\frac{1}{2}}\, \delta_z\big|_{i,j,k+\frac{1}{2}} \tag{7.55}
$$

for all edges with a non-zero fractional length. The permeability is also scaled by the fractional area. Namely, let

$$
\tilde{\mu}_x\big|_{i,j+\frac{1}{2},k+\frac{1}{2}} = \mu S_x\big|_{i,j+\frac{1}{2},k+\frac{1}{2}}, \quad \tilde{\mu}_y\big|_{i+\frac{1}{2},j,k+\frac{1}{2}} = \mu S_y\big|_{i+\frac{1}{2},j,k+\frac{1}{2}}, \quad \tilde{\mu}_z\big|_{i+\frac{1}{2},j+\frac{1}{2},k} = \mu S_z\big|_{i+\frac{1}{2},j+\frac{1}{2},k}. \tag{7.56}
$$

As a result, (7.52) is rewritten as:

$$
H_z^{n+1}\Big|_{i+\frac{1}{2},j+\frac{1}{2},k} = H_z^n\Big|_{i+\frac{1}{2},j+\frac{1}{2},k} - \frac{\Delta t}{\tilde{\mu}_z\big|_{i+\frac{1}{2},j+\frac{1}{2},k}}
$$
$$
\cdot \left(\frac{\tilde{E}_y^{n+\frac{1}{2}}\big|_{i+1,j+\frac{1}{2},k} - \tilde{E}_y^{n+\frac{1}{2}}\big|_{i,j+\frac{1}{2},k}}{\Delta x} - \frac{\tilde{E}_x^{n+\frac{1}{2}}\big|_{i+\frac{1}{2},j+1,k} - \tilde{E}_x^{n+\frac{1}{2}}\big|_{i+\frac{1}{2},j,k}}{\Delta y} \right), \tag{7.57}
$$

which resembles the standard FDTD update (3.27). Similar updates are derived for H_x and H_y. The scaling of the electric fields will also impact the dual electric field updates. For example, from (3.28):

$$
\tilde{E}_x^{n+\frac{1}{2}}\Big|_{i+\frac{1}{2},j,k} = \tilde{E}_x^{n-\frac{1}{2}}\Big|_{i+\frac{1}{2},j,k} + \frac{\Delta t}{\tilde{\varepsilon}_x\big|_{i+\frac{1}{2},j,k}} \left[\left(\frac{H_z^n\big|_{i+\frac{1}{2},j+\frac{1}{2},k} - H_z^n\big|_{i+\frac{1}{2},j-\frac{1}{2},k}}{\Delta y} \right) \right.
$$
$$
\left. - \left(\frac{H_y^n\big|_{i+\frac{1}{2},j,k+\frac{1}{2}} - H_y^n\big|_{i+\frac{1}{2},j,k-\frac{1}{2}}}{\Delta z} \right) \right] \tag{7.58}
$$

where

$$
\tilde{\varepsilon}_x\big|_{i+\frac{1}{2},j,k} = \varepsilon/\delta_x\big|_{i+\frac{1}{2},j,k}. \tag{7.59}
$$

Similar expressions are derived for $\tilde{E}_y$ and $\tilde{E}_z$.

After applying the scaling, the update equations for the conformal FDTD algorithm appear identical to those derived via the standard Yee-algorithm in (3.25)–(3.30). The difference is that the material constants are effectively scaled via the partial areas and lengths. Thus, the stability limit derived for the standard Yee-algorithm in Section 3.5 (c.f. (3.63)) again applies to this set of difference equations. The time stability limit of the conformal FDTD algorithm is expressed as:

$$
\Delta t = \frac{1}{\tilde{c}_{max}} \frac{\text{CFLN}}{\sqrt{\frac{1}{\Delta x^2} + \frac{1}{\Delta y^2} + \frac{1}{\Delta z^2}}} \tag{7.60}
$$

where CFLN is the Courant-Fredrichs-Lewy (CFL) number, which must be less than one, and

$$
\frac{1}{\tilde{c}_{max}} = \min\left(\sqrt{\tilde{\mu}\tilde{\varepsilon}} \right) \tag{7.61}
$$

where $\tilde{\mu}$, $\tilde{\varepsilon}$ are the normalized material constants of all deformed edge/face pairs. From (7.56) and (7.59), this can also be expressed as:

$$
\frac{1}{\tilde{c}_{max}} = \min\left(\sqrt{\frac{S}{\delta}} \sqrt{\mu\varepsilon} \right) \tag{7.62}
$$

where S/δ represents the set of all local face/edge pairs. Therefore, combining (7.62) with (3.63), the time-step of the DM-algorithm can be calculated as a fraction of the time-step of the base Yee-algorithm as:

$$\Delta t = \min \left(\sqrt{\frac{S}{\delta} \frac{c_{max}}{c_l}} \right) \cdot \Delta t^{Yee} \tag{7.63}$$

where S/δ represents the ratio of a partial face area and the partial edge length of an edge that bounds the face. Also, $c_l = 1/\sqrt{\mu_l \varepsilon_l}$ where μ_l and ε_l are the local permeability and permittivity of the face/edge pair, respectively. Finally, c_{max} is the maximum wave speed used to estimate the time-step Δt^{Yee} of the base uniform Yee-grid in (3.63). Because of this relationship, the smallest edge length of the deformed grid will govern the time-step. In order to keep the time-step from getting unreasonably small, Dey and Mittra suggest eliminating small faces by effectively absorbing them into the PEC.

7.3.2 YU-MITTRA (YM) CONFORMAL FDTD METHOD FOR CONDUCTING BOUNDARIES

One of the problems with the DM conformal FDTD algorithm is that the time-step can become extremely small if a fractional edge length outside of the PML becomes extremely small. Yu and Mittra proposed an alternative conformal algorithm that does not reduce the time-step of the underlying Yee-algorithm [13]. However, this comes at the expense of the loss of some accuracy. The Yu-Mittra (YM) conformal FDTD method is derived directly from the DM algorithm. The distinction is that the fractional area is always chosen to be one.

Again, consider the local deformation of a primary grid face due to a closed PEC cutting through the face as illustrated by Fig. 7.12. In the YM algorithm, Faraday's law is approximated as:

$$\frac{\partial}{\partial t} \mu H_z^{n+\frac{1}{2}} \Big|_{i+\frac{1}{2},j+\frac{1}{2},k} \Delta x \Delta y \approx$$

$$- \left(E_x^{n+\frac{1}{2}} \Big|_{i+\frac{1}{2},j,k} \Delta x + E_y^{n+\frac{1}{2}} \Big|_{i+1,j+\frac{1}{2},k} \ell_y \Big|_{i+1,j+\frac{1}{2},k} - E_x^{n+\frac{1}{2}} \Big|_{i+\frac{1}{2},j+1,k} \ell_x \Big|_{i+\frac{1}{2},j+1,k} - E_y^{n+\frac{1}{2}} \Big|_{i,j+\frac{1}{2},k} \Delta y \right). \tag{7.64}$$

Note that, in the YM-algorithm, the surface integral over the primary grid face is performed over the entire rectangular surface area. This is unlike the DM algorithm, which more accurately assumes the fractional area outside the PEC. The update equation is derived following the same procedure used in (7.51)–(7.58) with $A_z = \Delta x \Delta y$ and $S_z = 1$, the YM updates for H_z and E_x are expressed as:

$$H_z^{n+1} \Big|_{i+\frac{1}{2},j+\frac{1}{2},k} = H_z^n \Big|_{i+\frac{1}{2},j+\frac{1}{2},k} - \frac{\Delta t}{\mu} \left(\frac{\tilde{E}_y^{n+\frac{1}{2}} \Big|_{i+1,j+\frac{1}{2},k} - \tilde{E}_y^{n+\frac{1}{2}} \Big|_{i,j+\frac{1}{2},k}}{\Delta x} - \frac{\tilde{E}_x^{n+\frac{1}{2}} \Big|_{i+\frac{1}{2},j+1,k} - \tilde{E}_x^{n+\frac{1}{2}} \Big|_{i+\frac{1}{2},j,k}}{\Delta y} \right), \tag{7.65}$$

and

$$\tilde{E}_x^{n+\frac{1}{2}}\Big|_{i+\frac{1}{2},j,k} = \tilde{E}_x^{n-\frac{1}{2}}\Big|_{i+\frac{1}{2},j,k}$$
$$+ \frac{\Delta t}{\varepsilon_x\big|_{i+\frac{1}{2},j,k}}\left[\left(\frac{H_z^n\big|_{i+\frac{1}{2},j+\frac{1}{2},k} - H_z^n\big|_{i+\frac{1}{2},j-\frac{1}{2},k}}{\Delta y}\right) - \left(\frac{H_y^n\big|_{i+\frac{1}{2},j,k+\frac{1}{2}} - H_y^n\big|_{i+\frac{1}{2},j,k-\frac{1}{2}}}{\Delta z}\right)\right]. \qquad (7.66)$$

where the normalized electric field $\tilde{E}_x$ is defined by (7.55) and the local permittivity by (7.59).

The stability of the YM-algorithm is readily derived from (7.60)–(7.63). Since the YM algorithm assumes the fractional area $S = 1$, then from (7.63), $\Delta t^{YM} \leq \Delta t^{Yee}$ is guaranteed to be true. Therefore, the YM algorithm puts no further constraints on the time-step. The trade-off is a loss of accuracy.

7.3.3   BCK CONFORMAL FDTD METHOD FOR CONDUCTING BOUNDARIES

Benkler, Chavannes, and Kuster propose a variation of the DM conformal FDTD algorithm that allows the user to balance accuracy and stability [15]. We refer to this algorithm as the BCK conformal FDTD method.

The DM conformal FDTD algorithm is constrained by the time-stability limit of (7.63). Thus, the limit is constrained by the smallest ratio of S/δ. Interestingly, this ratio is driven by a small fractional area S, divided by the longest fractional length δ of the locally deformed face. This ratio is proportional to the smallest fractional edge length of the locally deformed face, which can become arbitrarily small for a general mesh. The BCK conformal FDTD algorithm suggests that the stability constraint can be loosened by increasing the fractional area, similar to the YM algorithm, rather than absorbing the cell in the PEC, as proposed by the DM algorithm. In this way, the accuracy can be 'tuned' to the desired CFL number (CFLN).

Let the user specify a minimum CFLN that constrains the time-step. The grid is then locally deformed and the partial lengths and areas are computed following the DM conformal FDTD algorithm. The time-step is then measured using (7.63) for every local face/edge pair. If any pair results in a time-step < the desired CFLN, the partial area of the face is increased until $\sqrt{S/\delta}(c_{max}/c_l) =$ CFLN. Thus, the tradeoff is to allow some local error in order to preserve the global time-step, and hence, maintain a reasonable computational time.

As an example consider the scattering of an electromagnetic wave by a PEC sphere computed with the proposed conformal FDTD schemes. The radius of the sphere is $\lambda/5.77$, where λ is the free-space wavelength. A uniform gridding of $\Delta x = \Delta y = \Delta z = \lambda/46.2$ was used. The mean square error (or L_2-error) was computed in the near field relative to the exact Mie-series solution [4]. This was computed for the uniform grid with a staircased mesh as well as conformal FDTD-grids based on the YM, the DM, and the BCK algorithms. The error versus the CFLN is presented in Fig. 7.13 (from [15]). Of course, the stair-case approximation leads to the largest error. The YM conformal-

FDTD algorithm leads to an improvement in the accuracy, and can still run at the Courant limit of the baseline Yee-algorithm. While the DM conformal FDTD algorithm improves the accuracy further, it must be run at a reduced time-step. Finally, the BCK algorithm can run at a variable time-step. As the CFLN is reduced, the accuracy is improved since the number of accurately deformed cells is increased. Nevertheless, at a CFLN = 1.0, it is more accurate than the YM-algorithm. For a CFLN = 0.5, the BCK algorithm leads to a full digit in improvement of the error as compared to the standard Yee-algorithm for this example. The error is also a factor of three smaller than the DM algorithm since all deformed cells are maintained in the locally conformal approximation.

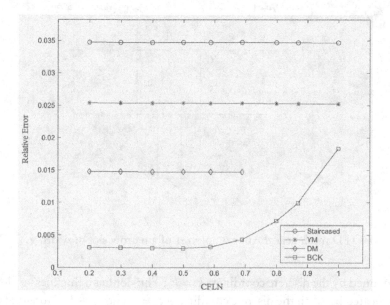

Figure 7.13: Near field error relative to the Mie-series solution for the scattering of a plane wave by a PEC sphere computed using the Yee-algorithm with a staircase approximation, and the YM, DM, and BCK conformal FDTD algoritms. *Source:* Benkler, et al., *IEEE Trans. Antennas and Propagation*, 2006, pp. 1843–1849. © 2006 IEEE.

7.4 NARROW SLOTS

Long, narrow slots in a conductor occurring at seams, joints, or air vents, have the potential to allow significant electromagnetic energy to penetrate through. In the field of Electromagnetic Compatibility (EMC), engineers find ways to reduce such coupling to reduce electromagnetic interference. On the other hand, antenna designers will use narrow slots as intentional radiators or couplers.

The challenge of modeling narrow slots is that their width can be much smaller than practical grid sizes required to discretized the super-structure. To accurately model the fields in and near the slot, one may be forced to reduce the global grid dimension. Alternatively, narrow-slot formulations have been developed that can accurately model the electromagnetic fields in the narrow slot without having to globally refine the grid [18]–[21]. The details of a narrow-slot formulation are provided in this section. In this formulation, the slot width is assumed to be small compared to the global cell size. The length of the slot is typically one or more grid cells. Also, the depth of the slot is also assumed to be one or more grid cells.

Consider a narrow slot in a thick PEC slab, as illustrated in Fig. 7.14. The upper aperture of

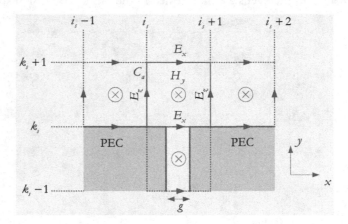

Figure 7.14: FDTD mesh near the upper aperture of a narrow slot of width g.

the slot is defined by the discrete coordinate $k = k_s$. The slot has a thickness g along the x-direction. The slot is located between the discrete coordinates $i = i_s$ and $i_s + 1$. (Note that this is not precisely defined.) The length of the slot is over the range $j \in j_{s_1}, j_{s_2}$. The local field update for the magnetic field in the cell just above the aperture of the slot can be applied via (7.52):

$$H_y^{n+1}\Big|_{i_s+\frac{1}{2},j,k_s+\frac{1}{2}} = H_y^n\Big|_{i_s+\frac{1}{2},j,k_s+\frac{1}{2}}$$
$$-\frac{\Delta t}{\mu}\left[\frac{E_x^n\Big|_{i_s+\frac{1}{2},j,k_s+1} - E_x^n\Big|_{i_s+\frac{1}{2},j,k_s}\ \delta x\Big|_{i_s+\frac{1}{2},j,k_s}}{\Delta z} - \frac{E_z^n\Big|_{i_s+1,j,k_s+\frac{1}{2}} - E_z^n\Big|_{i_s,j,k_s+\frac{1}{2}}}{\Delta x}\right]$$

$$(7.67)$$

where for this cell $S_y = 1$ and

$$\delta x\Big|_{i_s+\frac{1}{2},j,k_s} = \frac{g}{\Delta x}. \tag{7.68}$$

Next, consider a primary grid cell face that is inside of the narrow slot, as illustrated in Fig. 7.14. From (7.52):

$$H_{y}^{n+1}{}_{i_s+\frac{1}{2},j,k_s-\frac{1}{2}} = H_{y}^{n}{}_{i_s+\frac{1}{2},j,k_s-\frac{1}{2}}$$

$$- \frac{\Delta t}{\mu S_{y}{}_{i_s+\frac{1}{2},j,k_s-\frac{1}{2}}} \left[\frac{E_{x}^{n}{}_{i_s+\frac{1}{2},j,k_s} \delta_{x}{}_{i_s+\frac{1}{2},j,k_s} - E_{x}^{n}{}_{i_s+\frac{1}{2},j,k_s-1} \delta_{x}{}_{i_s+\frac{1}{2},j,k_s-1}}{\Delta z} \right]. \quad (7.69)$$

It is interesting to note that in this case, $\delta_{x}{}_{i_s+\frac{1}{2},j,k_s} = \delta_{x}{}_{i_s+\frac{1}{2},j,k_s-1} = S_{y}{}_{i_s+\frac{1}{2},j,k_s-\frac{1}{2}} = g/\Delta x$. Thus, this leads to the update:

$$H_{y}^{n+1}{}_{i_s+\frac{1}{2},j,k_s-\frac{1}{2}} = H_{y}^{n}{}_{i_s+\frac{1}{2},j,k_s-\frac{1}{2}} - \frac{\Delta t}{\mu} \left[\frac{E_{x}^{n}{}_{i_s+\frac{1}{2},j,k_s} - E_{x}^{n}{}_{i_s+\frac{1}{2},j,k_s-1}}{\Delta z} \right]. \quad (7.70)$$

Now, consider the electric field updates in the slot. From Fig. 7.14, it is observed that the secondary grid faces with centers defined by $E_{x}{}_{i_s+\frac{1}{2},j,k}$ will not be impacted. Consequently, the updates of these fields will be performed as defined by (3.28).

It is interesting to observe that the update of the axial magnetic field within the slot is independent of the slot width (c.f., (7.70)). Rather, the physical width of the slot is realized by the slot aperture fields above or below the slot, as governed by the updates (7.67). Also, only the lateral electric field and the axial magnetic field are modeled within the slot. The remaining fields are forced to be zero by the PEC.

Now, consider a slot with thickness less than a cell dimension, as illustrated in Fig. 7.15. The aperture of the slot lies in the plane $k = k_s$, and within the cell $i = i_s$. Following the previous procedure, it is observed that the update of the magnetic field in the cell above the slot ($H_{y}{}_{i_s+\frac{1}{2},j_s,k_s+\frac{1}{2}}$) will be identical to that in (7.67). The update of the magnetic field in the cell that contains the slot can be derived via (7.52) as:

$$H_{y}^{n+1}{}_{i_s+\frac{1}{2},j,k_s-\frac{1}{2}} = H_{y}^{n}{}_{i_s+\frac{1}{2},j,k_s-\frac{1}{2}} - \frac{\Delta t}{\mu S_{y}{}_{i_s+\frac{1}{2},j,k_s-\frac{1}{2}}} \cdot$$

$$\left[\frac{E_{x}^{n}{}_{i_s+\frac{1}{2},j,k_s} \delta_{x}{}_{i_s+\frac{1}{2},j,k_s} - E_{x}^{n}{}_{i_s+\frac{1}{2},j,k_s-1}}{\Delta z} - \frac{E_{z}^{n}{}_{i_s+1,j,k_s-\frac{1}{2}} \delta_{z}{}_{i_s+1,j,k_s-\frac{1}{2}} - E_{z}^{n}{}_{i_s,j,k_s-\frac{1}{2}} \delta_{z}{}_{i_s,j,k_s-\frac{1}{2}}}{\Delta x} \right]$$

$$(7.71)$$

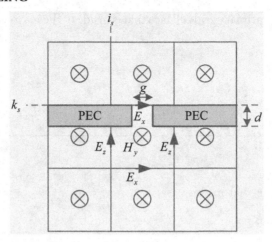

Figure 7.15: Narrow-slot of width g and thickness d, both of which are smaller than the grid dimensions.

where

$$S_{y}\bigg|_{i_s+\frac{1}{2},j,k_s-\frac{1}{2}} = \frac{gd + \Delta x(\Delta z - d)}{\Delta x \Delta z}, \quad \delta_{x}\bigg|_{i_s+\frac{1}{2},j,k_s} = \frac{g}{\Delta x},$$

$$\delta_{z}\bigg|_{i_s+1,j,k_s-\frac{1}{2}} = \delta_{z}\bigg|_{i_s,j,k_s-\frac{1}{2}} = \frac{\Delta z - d}{\Delta z}. \tag{7.72}$$

Following (7.54)–(7.59), this can be expressed in the form of a DM or BCK conformal-FDTD update equations. Also, note that all the cells beneath the PEC slab must also be deformed using the DM or BCK conformal-FDTD method.

7.5   CONFORMAL FDTD METHODS FOR MATERIAL BOUNDARIES

The previous two sections dealt with conformal FDTD methods for conducting boundaries. It is also desirable to establish an effective conformal FDTD method for penetrable material boundaries. In this section, we consider a dielectric boundary – that is, an arbitrary shaped boundary separating two regions of different dielectric constants. This formulation will also apply to finite conductors. One can derive a dual formulation for magnetic materials.

A number of methods have been proposed for conformal FDTD methods for dielectric boundaries [17],[22]–[28]. In this section, the method presented by Yu and Mittra (YM) [24] for isotropic dielectric boundaries is presented.

Consider the curved boundary separating two dielectric regions illustrated in Fig. 7.16. The boundary cuts across the primary grid face, and across two edges. In the YM conformal dielectric

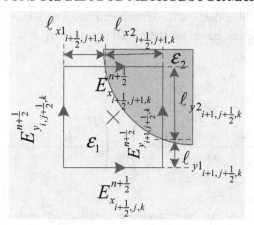

Figure 7.16: Curved boundary separating two dielectric regions cutting through a primary face FDTD cell.

FDTD method, the local permittivities associated with the electric field edges are modified using a local averaging scheme. Referring to Fig. 7.16, one would approximate:

$$\varepsilon_{x_{i+\frac{1}{2},j+1,k}} = \left(\ell_{x1_{i+\frac{1}{2},j+1,k}} \varepsilon_1 + \ell_{x2_{i+\frac{1}{2},j+1,k}} \varepsilon_2 \right) / \Delta x \tag{7.73}$$

$$\varepsilon_{y_{i+1,j+\frac{1}{2},k}} = \left(\ell_{y1_{i+1,j+\frac{1}{2},k}} \varepsilon_1 + \ell_{y2_{i+1,j+\frac{1}{2},k}} \varepsilon_2 \right) / \Delta y \tag{7.74}$$

where $\ell_{x2_{i+\frac{1}{2},j+1,k}} = \Delta x - \ell_{x1_{i+\frac{1}{2},j+1,k}}$ and $\ell_{y2_{i+1,j+\frac{1}{2},k}} = \Delta y - \ell_{y1_{i+1,j+\frac{1}{2},k}}$. Following the notation of (7.53),

$$\varepsilon_{x_{i+\frac{1}{2},j+1,k}} = \left(\delta_{x1_{i+\frac{1}{2},j+1,k}} \varepsilon_1 + \delta_{x2_{i+\frac{1}{2},j+1,k}} \varepsilon_2 \right) \tag{7.75}$$

$$\varepsilon_{y_{i+1,j+\frac{1}{2},k}} = \left(\delta_{y1_{i+1,j+\frac{1}{2},k}} \varepsilon_1 + \delta_{y2_{i+1,j+\frac{1}{2},k}} \varepsilon_2 \right). \tag{7.76}$$

Also, referring to Fig. 7.16, $\varepsilon_{x_{i+\frac{1}{2},j,k}} = \varepsilon_{y_{i,j+\frac{1}{2},k}} = \varepsilon_1$.

The magnetic field updates will be computed based on a normal Yee-update (c.f., (3.27)). The electric field updates will be based on their local permittivity. For example, from (7.73) and (3.28):

$$E_x^{n+\frac{1}{2}}\Big|_{i+\frac{1}{2},j+1,k} = E_x^{n-\frac{1}{2}}\Big|_{i+\frac{1}{2},j+1,k} + \frac{\Delta t}{\varepsilon_x\big|_{i+\frac{1}{2},j+1,k}}\cdot$$

$$\left[\left(\frac{H_z^n\big|_{i+\frac{1}{2},j+\frac{3}{2},k} - H_z^n\big|_{i+\frac{1}{2},j+\frac{1}{2},k}}{\Delta y}\right) - \left(\frac{H_y^n\big|_{i+\frac{1}{2},j+1,k+\frac{1}{2}} - H_y^n\big|_{i+\frac{1}{2},j+1,k-\frac{1}{2}}}{\Delta z}\right)\right]. \qquad (7.77)$$

A similar update is derived for E_y.

It is noted that this can be extended to conducting material. For example, let medium 1 and medium 2 have conductivities σ_1 and σ_2, respectively. Then, the local conductivity can be approximated as:

$$\sigma_x\big|_{i+\frac{1}{2},j+1,k} = \left(\delta_{x1}\big|_{i+\frac{1}{2},j+1,k}\,\sigma_1 + \delta_{x2}\big|_{i+\frac{1}{2},j+1,k}\,\sigma_2\right) \qquad (7.78)$$

$$\sigma_y\big|_{i+1,j+\frac{1}{2},k} = \left(\delta_{y1}\big|_{i+1,j+\frac{1}{2},k}\,\sigma_1 + \delta_{y2}\big|_{i+1,j+\frac{1}{2},k}\,\sigma_2\right). \qquad (7.79)$$

Consequently, the local updates are performed as defined by (3.99) with the local edge conductivity and permittivity.

As an example, consider the cylindrical dielectric resonator (DR) in a rectangular cavity as illustrated in Fig. 7.17. Three different dimensions were assumed, as detailed in the first column of Table 7.1 [24]. The dielectric resonator, with radius $2R$ and height t has a dielectric constant of 38.

Table 7.1: Comparison of different methods for computing the resonant frequency of a dielectric rod in a cavity. *Source:* Yu, et al., *IEEE Microwave and Wireless Component Letters,* 2001, pp. 25–27. © 2001 IEEE.

Size (inch)		Resonant Frequencies (GHz)			
2R	t	Yu-Mittra [24]	Kaneda's [25]	Mode Matching [29]	Measured [29]
0.654	0.218	4.388	4.40	4.3880	4.382
0.689	0.230	4.163	4.17	4.1605	4.153
0.757	0.253	3.725	3.78	3.721	3.777

The dielectric pedestal below the resonator has a dielectric constant of 1. Table 7.2 provides that the base FDTD algorithm assumes a non-uniform gridding, as detailed in Section 3.10. To better match the dimensions of the dielectric resonator (DR), the grid dimension in the DR is different than that outside. To model the curved boundaries of the DR, the Yu-Mittra conformal dielectric

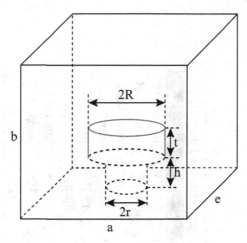

Figure 7.17: Cylindrical dielectric resonator in a rectangular box. *Source:* Yu, et al., *IEEE Microwave and Wireless Component Letters*, 2001, pp. 25–27. © 2001 IEEE.

FDTD scheme was used. The simulation was run for 40,000 time-steps. The results are compared to those computed via Kaneda, et al., [25] as well as Liang [29]. Measured data is also obtained from [29]. The Yu-Mittra algorithm provides an excellent approximation to the resonant frequency as compared to the mode matching method and the measured data.

7.6 THIN MATERIAL SHEETS

In this section, a subcell model for a thin-material sheet is presented [30, 31]. The distinction between this and the previous section is that the sheet is assumed to have a thickness less than the cell dimension. Such models could be necessary when modeling practical problems such as an antenna covered with a radome, or a waveguide loaded with resistive cards to suppress modes, or surfaces with thin film dielectric coatings. In the following presentation, the thin sheet is assumed to be a dielectric or conducting material. A dual model can be derived for a thin magnetic material sheet. For the subcell model to be valid, the thickness of the dielectric sheet should be small compared to a material wavelength. If the material is highly conductive, then the thickness of the sheet should be small compared to the skin depth. If these conditions hold true, the subcell model provides a very good approximation for the fields within and near the material sheet.

Consider a thin dielectric material sheet of thickness d with upper and lower surfaces with z-directed normals. A cross-section of the sheet is illustrated in Fig. 7.18. The thin sheet has an electric material profile $(\sigma_s, \varepsilon_s)$. It is assumed that d is $\ll$ the skin depth and the wavelength inside the sheet at the highest frequency of interest. The thickness d is also assumed to be less than the

Table 7.2: Dimensions of the non-uniform FDTD grid for the three geometries in Table 7.1.

Case	Grid. Dim.	Δx		Δy		Δz	
1	48 × 44 × 47	0.519112	mm in DR.	0.519112	mm in DR.	0.537307	mm below
		0.549275	mm outside	0.563033	mm outside	0.5537	mm in DR.
						0.536033	mm above
2	48 × 46 × 47	0.51472	mm in DR.	0.51472	mm in DR.	0.537308	mm below
		0.56424	mm outside	0.48895	mm outside	0.53109	mm in DR.
						0.54665	mm above
3	50 × 46 × 49	0.509947	mm in DR.	0.509947	mm in DR.	0.4989	mm below
		0.51435	mm outside	0.517525	mm outside	0.5355	mm in DR.
						0.5213	mm above

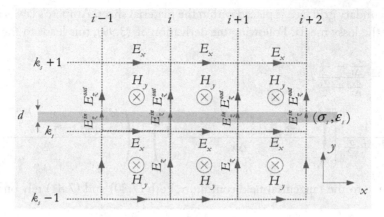

Figure 7.18: Thin material sheet of thickness d and electric material constants (σ_s, ε_s) within the primary FDTD grid.

grid dimension Δz. Relative to the discrete FDTD grid, the sheet is assumed to be located in the region above $z = k_s \Delta z$ and below $z = (k_s + 1) \Delta z$, as illustrated in Fig. 7.18.

The electric and magnetic fields that are tangential to the upper or lower surfaces of the thin sheet are continuous across these boundaries. Since the sheet is assumed to be very thin, it is justifiable to approximate that the tangential fields are continuous through the sheet. In this example, this would include E_x, E_y, H_x, and H_y. On the other hand, the normal electric field is discontinuous across the sheet. To accommodate this discontinuity, the normal electric field E_z is given separate degrees of freedom inside and outside of the sheet. The field E_z^{in} is assumed to be constant within the sheet. Similarly, E_z^{out} is assumed to be constant outside the sheet.

The electric field exterior to the sheet is evaluated via a discretization of Ampere's law over a secondary grid face that is located outside of the thin sheet. This leads to the expected update:

$$
E_z^{out}{}^{n+\frac{1}{2}}_{i,j,k_s+\frac{1}{2}} = E_z^{out}{}^{n-\frac{1}{2}}_{i,j,k_s+\frac{1}{2}}
$$

$$
+ \frac{\Delta t}{\varepsilon} \left[\left(\frac{H_y^n{}_{i+\frac{1}{2},j,k_s+\frac{1}{2}} - H_y^n{}_{i-\frac{1}{2},j,k_s+\frac{1}{2}}}{\Delta x} \right) - \left(\frac{H_x^n{}_{i,j+\frac{1}{2},k_s+\frac{1}{2}} - H_x^n{}_{i,j-\frac{1}{2},k_s+\frac{1}{2}}}{\Delta y} \right) \right].
$$

$$(7.80)$$

Another secondary grid face is placed within the material sheet. Ampere's law is again enforced over this face in the lossy media. Following the derivation of (3.99), this leads to the update:

$$
E_z^{in}{}^{n+\frac{1}{2}}_{i,j,ks+\frac{1}{2}} = \frac{\frac{\varepsilon_s}{\Delta t} - \frac{\sigma_s}{2}}{\frac{\varepsilon_s}{\Delta t} + \frac{\sigma_s}{2}} E_z^{in}{}^{n-\frac{1}{2}}_{i,j,ks+\frac{1}{2}}
$$
$$
+ \frac{1}{\frac{\varepsilon_s}{\Delta t} + \frac{\sigma_s}{2}} \left[\left(\frac{H_y^n{}_{i+\frac{1}{2},j,ks+\frac{1}{2}} - H_y^n{}_{i-\frac{1}{2},j,ks+\frac{1}{2}}}{\Delta x} \right) - \left(\frac{H_x^n{}_{i,j+\frac{1}{2},ks+\frac{1}{2}} - H_x^n{}_{i,j-\frac{1}{2},ks+\frac{1}{2}}}{\Delta y} \right) \right]. \quad (7.81)
$$

Note that due to the tangential field continuity, both (7.80) and (7.81) rely on the same magnetic fields.

The updates of the tangential electric fields E_x and E_y, in the vicinity of the thin sheet, also need special treatment. These updates are evaluated via a discretization of Ampere's law over secondary grid faces that cut across the inhomogeneous lossy dielectric media. This is performed following the scheme presented in Section 3.7. This leads to the update equations:

$$
E_x^{n+\frac{1}{2}}_{i+\frac{1}{2},j,ks} = \frac{\frac{\varepsilon_{ave}}{\Delta t} - \frac{\sigma_{ave}}{2}}{\frac{\varepsilon_{ave}}{\Delta t} + \frac{\sigma_{ave}}{2}} E_x^{n-\frac{1}{2}}_{i+\frac{1}{2},j,ks}
$$
$$
+ \frac{1}{\frac{\varepsilon_{ave}}{\Delta t} + \frac{\sigma_{ave}}{2}} \left(\frac{H_z^n{}_{i+\frac{1}{2},j+\frac{1}{2},ks} - H_z^n{}_{i+\frac{1}{2},j-\frac{1}{2},ks}}{\Delta y} \right) - \left(\frac{H_y^n{}_{i+\frac{1}{2},j,ks+\frac{1}{2}} - H_y^n{}_{i+\frac{1}{2},j,ks-\frac{1}{2}}}{\Delta z} \right) \quad (7.82)
$$

and

$$
E_y^{n+\frac{1}{2}}_{i,j+\frac{1}{2},ks} = \frac{\frac{\varepsilon_{ave}}{\Delta t} - \frac{\sigma_{ave}}{2}}{\frac{\varepsilon_{ave}}{\Delta t} + \frac{\sigma_{ave}}{2}} E_y^{n-\frac{1}{2}}_{i,j+\frac{1}{2},ks}
$$
$$
+ \frac{1}{\frac{\varepsilon_{ave}}{\Delta t} + \frac{\sigma_{ave}}{2}} \left[\left(\frac{H_x^n{}_{i,j+\frac{1}{2},ks+\frac{1}{2}} - H_x^n{}_{i,j+\frac{1}{2},ks-\frac{1}{2}}}{\Delta z} \right) - \left(\frac{H_z^n{}_{i+\frac{1}{2},j+\frac{1}{2},ks} - H_z^n{}_{i-\frac{1}{2},j+\frac{1}{2},ks}}{\Delta x} \right) \right]
$$
$$
(7.83)
$$

where,

$$
\varepsilon_{ave} = \left(1 - \frac{d}{\Delta z} \right) \varepsilon + \frac{d}{\Delta z} \varepsilon_s, \qquad \sigma_{ave} = \frac{d}{\Delta z} \sigma_s. \quad (7.84)
$$

Next, consider the updates of the tangential magnetic fields H_x and H_y in the vicinity of the thin material sheet. These are derived from the discretization of Faraday's law performed about the primary grid faces that cut across the thin-material sheet. Note that the line integrals over the z-directed edges involve both E_z^{in} and E_z^{out}. The updates for the tangential magnetic fields are then expressed as:

$$
H_{x_{i,j+\frac{1}{2},ks+\frac{1}{2}}}^{n+1} = H_{x_{i,j+\frac{1}{2},ks+\frac{1}{2}}}^{n} + \frac{\Delta t}{\mu} \left(\frac{E_{y_{i,j+\frac{1}{2},ks+1}}^{n+\frac{1}{2}} - E_{y_{i,j+\frac{1}{2},ks}}^{n+\frac{1}{2}}}{\Delta z} \right) -
$$

$$
\frac{\Delta t}{\mu} \left[\frac{\Delta z - d}{\Delta z} \left(\frac{E_{z_{i,j+1,ks+\frac{1}{2}}}^{out \ n+\frac{1}{2}} - E_{z_{i,j,ks+\frac{1}{2}}}^{out \ n+\frac{1}{2}}}{\Delta y} \right) + \frac{d}{\Delta z} \left(\frac{E_{z_{i,j+1,ks+\frac{1}{2}}}^{in \ n+\frac{1}{2}} - E_{z_{i,j,ks+\frac{1}{2}}}^{in \ n+\frac{1}{2}}}{\Delta y} \right) \right] \tag{7.85}
$$

and

$$
H_{y_{i+\frac{1}{2},j,k+\frac{1}{2}}}^{n+1} = H_{y_{i+\frac{1}{2},j,k+\frac{1}{2}}}^{n} - \frac{\Delta t}{\mu} \left(\frac{E_{x_{i+\frac{1}{2},j,k+1}}^{n+\frac{1}{2}} - E_{x_{i+\frac{1}{2},j,k}}^{n+\frac{1}{2}}}{\Delta z} \right) +
$$

$$
\frac{\Delta t}{\mu} \left[\frac{\Delta z - d}{\Delta z} \left(\frac{E_{z_{i+1,j,k+\frac{1}{2}}}^{out \ n+\frac{1}{2}} - E_{z_{i,j,k+\frac{1}{2}}}^{out \ n+\frac{1}{2}}}{\Delta x} \right) + \frac{d}{\Delta z} \left(\frac{E_{z_{i+1,j,k+\frac{1}{2}}}^{in \ n+\frac{1}{2}} - E_{z_{i,j,k+\frac{1}{2}}}^{in \ n+\frac{1}{2}}}{\Delta x} \right) \right]. \tag{7.86}
$$

The field updates in (7.80)–(7.86) are the complete updates for the z-normal thin material sheet. Similar updates can be derived for other orientations. Also, duality can be applied to derive updates for a magnetic material sheet. The update equations are stable within the normal Courant limit, and thus provide a good approximation for thin material sheets without significantly increasing the resources needed to perform the global FDTD simulation.

7.7   PROBLEMS

1. Derive Equation (7.4) from (7.3) by assuming the local singular field approximations of (7.1) and (7.2).

2. Derive Equations (7.52), (7.57), and (7.58).

3. Following the derivation of field update equations for inhomogeneous media in Section 3.7, derive Equations (7.82)–(7.84).

4. Derive (7.85) and (7.86).

5. Derive normalizations for the thin wire algorithm presented in Sections 7.2.2 and 7.2.3 to conform to the unified FDTD update expressions of Section B.6 of Appendix B.

6. Derive the DM-conformal FDTD algorithm for a closed PEC body (Section 7.3) for the case when the background media is lossy.

7. Derive the YM-conformal FDTD algorithm for a dielectric body for a lossy dielectric with conductivity σ.

REFERENCES

[1] C. A. Balanis, *Advanced Engineering Electromagnetics*. New York: Wiley, 1989. 138

[2] K. R. Umashankar, A. Taflove, and B. Beker, "Calcuation and experimental validation of induced currents on coupled wires in an arbitrary shaped cavity," *IEEE Transactions on Antennas and Propagation*, vol. 35, pp. 1248–1257, Nov 1987. DOI: 10.1109/TAP.1987.1144000 138

[3] R. M. Mäkinen, J. S. Juntunen, and M. A. Kivikoski, "An improved thin-wire model for FDTD," *IEEE Transactions on Microwave Theory and Techniques*, vol. 50, pp. 1245–1255, May 2002. DOI: 10.1109/22.999136 139, 140, 144, 146, 147

[4] J. A. Stratton, *Electromagnetic Theory*. New York: McGraw-Hill, 1941. 143, 158

[5] A. R. Krommer and C. W. Ueberhuber, *Computational Integration*. Philadelphia: SIAM, 1998. 145, 146

[6] S. Zhang and J. M. Jin, *Computation of Special Functions*. NY: John Wiley & Sons, Inc., 1996. 145

[7] S. Watanabe and M. Taki, "An improved FDTD model for the feeding gap of a thin-wire antenna," *IEEE Microwave and Guided Wave Letters*, vol. 8, pp. 152–154, Apr 1998. DOI: 10.1109/75.663515 146

[8] G. J. Burke and A. J. Poggio, "Numerical Electromagnetics Code (NEC)-method of moments," Naval Ocean Systems Center, San Diego, CA Technical Document 11, January 1981. 149

[9] J. G. Maloney, K. L. Shlager, and G. S. Smith, "A Simple FDTD Model for Transient Excitation of Antennas by Transmission-Lines," *IEEE Transactions on Antennas and Propagation*, vol. 42, pp. 289–292, Feb 1994. DOI: 10.1109/8.277228 150

[10] T. G. Jurgens, A. Taflove, K. R. Umashankar, and T. G. Moore, "Finite-difference time-domain modeling of curved surfaces," *IEEE Transactions on Antennas and Propagation*, vol. 40, pp. 357–366, 1992. DOI: 10.1109/8.138836 153

[11] S. Dey and R. Mittra, "A locally conformal finite-difference time-domain (FDTD) algorithm for modeling three-dimensional perfectly conducting objects," *IEEE Microwave And Guided Wave Letters*, vol. 7, pp. 273–275, Sep 1997. DOI: 10.1109/75.622536 153

[12] S. Dey and R. Mittra, "A modified locally conformal finite-difference time-domain algorithm for modeling three-dimensional perfectly conducting objects," *Microwave and Optical Technology Letters*, vol. 17, pp. 349–352, Apr 1998.
DOI: 10.1002/(SICI)1098-2760(19980420)17:6%3C349::AID-MOP4%3E3.0.CO;2-H
153

[13] W. H. Yu and R. Mittra, "A conformal FDTD software package modeling antennas and microstrip circuit components," *IEEE Antennas and Propagation Magazine*, vol. 42, pp. 28–39, Oct 2000. DOI: 10.1109/74.883505 153, 157

[14] G. Waldschmidt and A. Taflove, "Three-dimensional CAD-based mesh generator for the Dey-Mittra conformal FDTD algorithm," *IEEE Transactions on Antennas and Propagation*, vol. 52, pp. 1658–1664, Jul 2004. DOI: 10.1109/TAP.2004.831334

[15] S. Benkler, N. Chavannes, and N. Kuster, "A new 3-D conformal PEC FDTD scheme with user-defined geometric precision and derived stability criterion," *IEEE Transactions on Antennas and Propagation*, vol. 54, pp. 1843–1849, Jun 2006. DOI: 10.1109/TAP.2006.875909 153, 158

[16] T. Xiao and Q. H. Liu, "A 3-D enlarged cell technique (ECT) for the conformal FDTD method," *IEEE Transactions on Antennas and Propagation*, vol. 56, pp. 765–773, Mar 2008. DOI: 10.1109/TAP.2008.916876

[17] W. H. Yu, R. Mittra, X. L. Yang, Y. J. Liu, Q. J. Rao, and A. Muto, "High-Performance Conformal FDTD Techniques," *IEEE Microwave Magazine*, vol. 11, pp. 42–55, Jun 2010. DOI: 10.1109/MMM.2010.936496 153, 162

[18] J. Gilbert and R. Holland, "Implementation of the thin-slot formalism in the finite difference EMP code THREADII," *IEEE Transactions on Nuclear Science*, vol. 28, pp. 4269–4274, 1981. DOI: 10.1109/TNS.1981.4335711 160

[19] K. P. Ma, M. Li, J. L. Drewniak, T. H. Hubing, and T. P. VanDoren, "Comparison of FDTD algorithms for subcellular modeling of slots in shielding enclosures," *IEEE Transactions on Electromagnetic Compatibility*, vol. 39, pp. 147–155, May 1997. DOI: 10.1109/15.584937

[20] A. Taflove, K. R. Umashankar, B. Beker, F. Harfoush, and K. S. Yee, "Detailed FD-TD analysis of electromagnetic-fields penetrating narrow slots and lapped joints in thick conducting screens," *IEEE Transactions on Antennas and Propagation*, vol. 36, pp. 247–257, Feb 1988. DOI: 10.1109/8.1102

[21] C. D. Turner and L. D. Bacon, "Evaluation of a thin-slot formalism for finite-difference time-domain electromagnetic codes," *IEEE Transactions on Electromagnetic Compatibility*, vol. 30, pp. 523–528, 1988. DOI: 10.1109/15.8766 160

[22] S. Dey and R. Mittra, "A conformal finite-difference time-domain technique for modeling cylindrical dielectric resonators," *IEEE Transactions on Microwave Theory and Techniques*, vol. 47, pp. 1737–1739, Sep 1999. DOI: 10.1109/22.788616 162

[23] D. Li, P. M. Meaney, and K. D. Paulsen, "Conformal microwave imaging for breast cancer detection," *IEEE Transactions On Microwave Theory And Techniques*, vol. 51, pp. 1179–1186, Apr 2003. DOI: 10.1109/TMTT.2003.809624

[24] W. H. Yu and R. Mittra, "A conformal finite difference time domain technique for modeling curved dielectric surfaces," *IEEE Microwave and Wireless Components Letters*, vol. 11, pp. 25–27, Jan 2001. DOI: 10.1109/7260.905957 162, 164

[25] N. Kaneda, B. Houshmand, and T. Itoh, "FDTD analysis of dielectric resonators with curved surfaces," *IEEE Transactions on Microwave Theory and Techniques*, vol. 45, pp. 1645–1649, 1997. DOI: 10.1109/22.622937 164, 165

[26] X. J. Hu and D. B. Ge, "Study on conformal FDTD for electromagnetic scattering by targets with thin coating," *Progress in Electromagnetics Research-Pier*, vol. 79, pp. 305–319, 2008. DOI: 10.2528/PIER07101902

[27] T. I. Kosmanis and T. D. Tsiboukis, "A systematic Conformal finite-difference time-domain (FDTD) technique for the simulation of arbitrarily curved interfaces between dielectrics," *IEEE Transactions on Magnetics*, vol. 38, pp. 645–648, Mar 2002. DOI: 10.1109/20.996168

[28] G. R. Werner and J. R. Cary, "A stable FDTD algorithm for non-diagonal, anisotropic dielectrics," *Journal of Computational Physics*, vol. 226, pp. 1085–1101, Sep 2007. DOI: 10.1016/j.jcp.2007.05.008 162

[29] X.-P. Liang and K. A. Zakim, "Modeling of cylindrical dielectric resonators in rectangular waveguides and cavity," *IEEE Transactions on Microwave Theory and Techniques*, vol. 41, pp. 2174–2181, 1993. DOI: 10.1109/22.260703 164, 165

[30] J. G. Maloney and G. S. Smith, "The efficient modeling of thin matieral sheets in the finite-difference time-domain (FDTD) method," *IEEE Transactions on Antennas and Propagation*, vol. 40, pp. 323–330, Mar 1992. DOI: 10.1109/8.135475 165

[31] J. G. Maloney and G. S. Smith, "A comparison of methods for modeling electrically thin dielectric and conducting sheets in the finite-difference time-domain method," *IEEE Transactions on Antennas and Propagation*, vol. 41, pp. 690–694, May 1993. DOI: 10.1109/8.222291 165

CHAPTER 8

Post Processing

8.1 INTRODUCTION

When the FDTD method is used to solve the time-dependent Maxwell's equations, measurable quantities such as a port voltage or current, an input impedance, a network parameterization, a radiation pattern, or a specific absorption rate are often required. Such quantities are indirect measurements that are computed directly from the time-dependent electromagnetic fields. Such entities are typically necessary to fully characterize a device under test, and also enable the device to be part of the design of a broader system.

The purpose of this chapter is to discuss a few practical post-processing methods that can be derived from the FDTD method. The methods include discrete port parameterizations, network analysis methods, and near-field to far field transforms. While this set is not exhaustive, it should provide a basic foundation to enable the reader to understand and develop a broader range of post processing methods.

8.2 NETWORK ANALYSIS

Network parameters provide a reduced order model of a more complex linear or non-linear device [1]. Network parameters establish relationships between port quantities, such as port voltages and currents or inward and outward traveling waves. A device can be modeled within a more complex system using only its network parameters, rather than having to represent the full detail of the device itself. The FDTD method can be used to accurately and efficiently extract the network parameters of complex linear and non-linear devices.

In this section, the measuring of the network parameterization of multiport devices via the FDTD method is presented. The focus is on devices with terminals that support either transverse electromagnetic (TEM) waves, or quasi-TEM waves. For this class of devices, the ports are characterized by a port voltage and current. Section 8.2.1 introduces methods to compute the port voltage, current, and characteristic impedance. The following two sections discuss how the FDTD method can be used to extract the admittance and scattering parameters, respectively, of a multiport device.

8.2.1 DISCRETE NETWORK PORT PARAMETERIZATION

A discrete network port can be parameterized via a port voltage, current and impedance. In a FDTD simulation, such quantities are typically computed using quasi-static approximations. A voltage is typically associated with either a TEM or quasi-TEM wave supported by two or more conductors.

The voltage between the conductors is estimated using a quasi-static approximation based on the line integral:

$$V(t) = -\int_{C_V} \vec{E}(\vec{r}, t) \cdot d\vec{\ell} \qquad (8.1)$$

where C_V is a path between two conductors. For example, consider the cross-section of a uniform microstrip line illustrated in Fig. 8.1. The microstrip line supports a quasi-TEM wave propagating

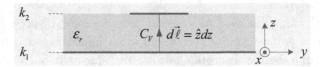

Figure 8.1: Cross-section of a microstrip line. C_V is the path of integration used for the voltage calculation.

along the x-axis. The line voltage at a position x can be estimated via a path integral between the ground plane and the microstrip. The path must lie in the cross-sectional plane. Typically, the shortest distance between the two conductors is chosen, as illustrated in Fig. 8.1. For this path, the discrete line voltage is estimated as:

$$V_i^{n+\frac{1}{2}} = -\sum_{k=k_1}^{k_2-1} E_z \Big|_{i, j_c, k+\frac{1}{2}}^{n+\frac{1}{2}} \Delta z \qquad (8.2)$$

where the ground plate is located at discrete index k_1 and the microstrip at k_2. The path is computed at the line center at discrete coordinate $j = j_c$. The line voltage is computed at the discrete x-coordinate i.

The line current is also extracted from a FDTD simulation using a quasi-static approximation of Ampere's law

$$I(t) = \oint_{C_I} \vec{H} \cdot d\vec{\ell} \qquad (8.3)$$

where C_I is a closed contour that encloses the conductor through which the net current, I, flows. The contour of integration is illustrated in Fig. 8.2. The direction of current flow is determined from the direction of the closed contour integral and using the right-hand rule. Using the contour in Fig. 8.2, the current will be measured flowing in the positive x-direction. The time-dependent current is exact if C_I actually touches the conductor surface. However, in the discrete FDTD space, the magnetic field tangential to the conductor surface is one-half of a cell away from the conductor on all sides (c.f., Fig. 8.3). As a consequence, (8.3) is an approximation of Ampere's law since it neglects the time rate of decrease of any electric flux cutting through the contour.

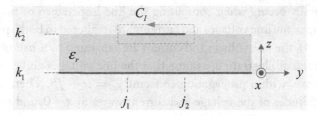

Figure 8.2: Cross-section of a microstrip line. C_I is the path of integration for the current calculation.

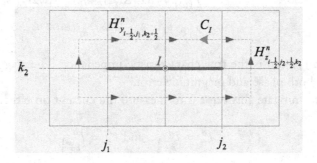

Figure 8.3: Contour C_I defined within the discrete FDTD grid.

Referring to Fig. 8.3, the microstrip lies in the $k = k_2$ plane and the cross section of the strip lies over the range $j \in j_1..j_2$. The net current flowing in the microstrip in the positive x-direction is approximated from the discrete FDTD fields as:

$$I^n_{i-\frac{1}{2}} = \sum_{j=j_1}^{j_2} \left(H^n_{y_{i-\frac{1}{2},j,k_2-\frac{1}{2}}} - H^n_{y_{i-\frac{1}{2},j,k_2+\frac{1}{2}}} \right) \Delta y$$

$$+ \left(H^n_{z_{i-\frac{1}{2},j_2+\frac{1}{2},k_2}} - H^n_{z_{i-\frac{1}{2},j_1-\frac{1}{2},k_2}} \right) \Delta z. \tag{8.4}$$

Note that the current is actually measured at $i - \frac{1}{2}$ due to the staggering of the fields.

Another useful quantity to extract for the uniform microstrip line port is the line impedance. The line impedance is computed in the frequency domain as:

$$Z(x, \omega) = \frac{V(x, \omega)}{I(x, \omega)} \tag{8.5}$$

where $\omega = 2\pi f$ is the radial frequency. The frequency dependent quantities, $V(x, \omega)$ and $I(x, \omega)$, are computed using a Fourier transform of the time-dependent quantities. Since the $V(x, t)$ and $I(x, t)$ are stored using uniformly distributed discrete samples in time, the Fourier transform is very conveniently computed via the Fast Fourier Transform (FFT) [2].

Error naturally occurs when computing the line impedance of the microstrip line with the FDTD method since the line voltages and currents are staggered in both space and time. As observed from (8.2) and (8.4), the line voltage and current are separated by a half of a space-cell and a half of a time-step. For sake of illustration, assume that the line voltages and currents represent a forward traveling TEM wave, with a propagation constant $\gamma_x = \alpha + j\beta$. Then, if we let $V(\omega)$ and $I(\omega)$ represent the amplitudes of the voltage and current waves at $x = 0$, and $t = 0$, the impedance in the Fourier domain is estimated as

$$\tilde{Z}(x, \omega) = \frac{V(\omega) e^{-\alpha x} e^{-j\beta x} e^{j\omega(t+\Delta t/2)}}{I(\omega) e^{-\alpha(x-\Delta x/2)} e^{-j\beta(x-\Delta x/2)} e^{j\omega t}}$$

$$= Z_0 e^{-\alpha \Delta x/2} e^{-j\beta \Delta x/2} e^{j\omega \Delta t/2} \tag{8.6}$$

where $Z_0 = V(\omega)/I(\omega)$ is the true characteristic impedance. The staggering of the voltage and current results in both phase and magnitude error.

One way to mitigate this error is to measure the current on either side of the line voltage. That is, let

$$I_-^n = I_{i-\frac{1}{2}}^n, \quad \text{and } I_+^n = I_{i+\frac{1}{2}}^n \tag{8.7}$$

which are computed using (8.4). The line impedance is then estimated using the geometric mean of these currents [3]:

$$Z(x, \omega) = \frac{V(x, \omega) e^{-j\omega \Delta t/2}}{\sqrt{I_-(x, \omega) I_+(x, \omega)}}. \tag{8.8}$$

Following a similar analogy as (8.6), the line impedance is evaluated as

$$\tilde{Z}(x, \omega) = \frac{V(\omega) e^{-\alpha x} e^{-j\beta x} e^{j\omega(t+\Delta t/2)} \cdot e^{-j\omega \Delta t/2}}{\sqrt{I(\omega) e^{-\alpha(x-\Delta x/2)} e^{-j\beta(x-\Delta x/2)} e^{j\omega t} \cdot I(\omega) e^{-\alpha(x+\Delta x/2)} e^{-j\beta(x+\Delta x/2)} e^{j\omega t}}}$$

$$= \frac{V(\omega)}{I(\omega)} = Z_0 \tag{8.9}$$

which is the desired characteristic line impedance without the delay errors.

The line impedance calculation correction of (8.9) is necessary when the current and voltage are separated by $1/2$ of a cell in both space and time. This occurs when the line integral for the voltage is perpendicular to the direction of current flow. However, if the line integral for the voltage is parallel to the current flow, the port voltage and current will be co-located in space. One example of this would be the measure of the input impedance of a dipole antenna. Consider the thin-wire dipole antenna excited by a delta-gap source, as illustrated in Fig. 7.5. The delta-gap has a voltage

$$V_{k_{gap}+\frac{1}{2}}^{n+\frac{1}{2}} = V_\delta\left(\left(n + \frac{1}{2}\right)\Delta t\right) \tag{8.10}$$

and is located effectively at the edge center defining the source gap as illustrated in Fig. 7.5. The current through the delta-gap source is approximated via the line integral of the magnetic fields about the delta gap. The current density is then approximated via (8.3) as:

$$
I^n_{k_{gap}+\frac{1}{2}} = \Delta y h_x \left(\tilde{H}^n_{y}\Big|_{i_w+\frac{1}{2},j_w,k_{gap}+\frac{1}{2}} - \tilde{H}^n_{y}\Big|_{i_w-\frac{1}{2},j_w,k_{gap}+\frac{1}{2}} \right)
$$
$$
- \Delta x h_y \left(\tilde{H}^n_{x}\Big|_{i_w,j_w+\frac{1}{2},k_{gap}+\frac{1}{2}} - \tilde{H}^n_{y}\Big|_{i_w,j_w-\frac{1}{2},k_{gap}+\frac{1}{2}} \right) \tag{8.11}
$$

where $\tilde{H}_y$ is defined by (7.1), and h_x and h_y are defined by (7.12) and (7.18), respectively. Once again, the voltage and current can be cast into the frequency domain via the FFT. Consequently, the input impedance of the thin dipole can be estimated as:

$$
Z_{in}(\omega) = \frac{V_{k_{gap}+\frac{1}{2}}(\omega)\, e^{-j\omega\Delta t/2}}{I_{k_{gap}+\frac{1}{2}}(\omega)} \tag{8.12}
$$

where the phase shift in the numerator is added to cancel the phase difference between the voltages and currents, which are separated by one-half of a time-step.

8.2.2 ADMITTANCE-PARAMETERS

The admittance (Y) parameters provide a direct relationship between the port currents and the port voltages of a multiport device. Let V_i be the frequency dependent voltage of the i-th port, and I_i be the frequency dependent port current. The Y-parameterization for the N-port device is expressed as:

$$
\begin{pmatrix} I_1 \\ I_2 \\ \vdots \\ I_N \end{pmatrix} = \begin{pmatrix} Y_{11} & Y_{12} & \cdots & Y_{1N} \\ Y_{21} & Y_{22} & \cdots & Y_{2N} \\ \vdots & \vdots & \ddots & \vdots \\ Y_{N1} & Y_{N2} & \cdots & Y_{NN} \end{pmatrix} \begin{pmatrix} V_1 \\ V_2 \\ \vdots \\ V_N \end{pmatrix} \tag{8.13}
$$

where $Y_{i,j}$ is the Y-parameter coupling the i-th and j-th ports. The Y-parameters can be extracted from a device by exciting the j-th port with a voltage, V_j, and terminating all the other ports with a short circuit, which restricts the voltage of these ports to 0. The port currents on all ports are then measured, thus allowing the computation of the j-th column of the admittance matrix from the quotient:

$$
Y_{i,j} = \frac{I_i}{V_j}\Bigg|_{\substack{V_i=0, \\ i \neq j}}. \tag{8.14}
$$

In this way, the Y-parameters are sometimes referred to as the short-circuit admittance parameters.

The admittance parameters can be extracted from a device via the FDTD method following this exact procedure. For example, consider an N-port device. The Y-parameters of the N-port can be extracted by exciting one of the ports with an ideal voltage source (with zero internal resistance). All the other ports are terminated with a short circuit. The port currents are computed from the FDTD simulation using (8.4) for all N ports. The currents are Fourier transformed, and the j-th column of the admittance matrix is computed using (8.14). If the port currents are staggered in space and time relative to the port voltages, then the Y-parameters are computed as:

$$Y_{i,j} = \left. \frac{\sqrt{I_i^- I_i^+} e^{j\omega \Delta t/2}}{V_j} \right|_{\substack{V_i=0, \\ i \neq j}}, \tag{8.15}$$

similar to (8.8). If the port voltage and current are co-located in space and only staggered in time, then $I_i^- = I_i^+ = I_i$.

To extract the full Y-matrix, N-such simulations must be performed exciting one port at a time. If there are geometric symmetries, this can be taken advantage of to reduce the number of simulations. It is also noted that if the device being modeled is reciprocal (passive, linear, and isotropic media), then the admittance matrix will be symmetric [1].

A word of caution is necessary for computing the Y-parameters. In order to properly compute the port currents, the system under test must be simulated to a steady state. This typically means that the port currents settle to a zero value. If the device being modeled has very low loss (including radiation losses), then reaching a steady state can require a very long time simulation. One way this can be mitigated is to extrapolate the time-domain port quantities based on a FDTD simulation over a reduced time interval [4]–[6]. If possible, another alternative is to extract the scattering parameters of the device using matched load terminations as discussed in the next section.

8.2.3 SCATTERING PARAMETERS

The scattering parameters are quite different than the Y-parameters. Rather than providing a relationship of port voltages and currents, the scattering parameters provide relationships between inward and outward traveling waves of a multi-port device. The scattering parameters also quantify the flow of power through a multiport network [7]. As an example, consider an N-port network where each port supports a TEM wave. All ports are initially assumed to have the same characteristic impedance. The line voltage at each port is represented as a superposition of traveling waves. The traveling waves are assumed to be in the frequency domain with an $e^{j\omega t}$ time dependence. Let V_i^+ represent the *inward* traveling wave of port i. That is, it is a forward traveling wave flowing into the network. Let V_i^- represent the *outward* traveling wave of port i, which is a backward traveling flowing out of the network. The total port voltage is then represented as a superposition of both the inward and the outward traveling waves:

$$V_i = V_i^+ + V_i^-. \tag{8.16}$$

The scattering parameters define the relationship between the inward and outward traveling waves of the network:

$$
\begin{pmatrix} V_1^- \\ V_2^- \\ \vdots \\ V_N^- \end{pmatrix} = \begin{pmatrix} S_{11} & S_{12} & \cdots & S_{1N} \\ S_{21} & S_{22} & \cdots & S_{2N} \\ \vdots & \vdots & \ddots & \vdots \\ S_{N1} & S_{N2} & \cdots & S_{NN} \end{pmatrix} \begin{pmatrix} V_1^+ \\ V_2^+ \\ \vdots \\ V_N^+ \end{pmatrix}.
$$
(8.17)

The matrix is referred to as the scattering matrix, or the S-matrix, of the network.

The scattering parameters of a multiport network can be measured as follows. The j-th port is excited by a *matched* source, launching an inward traveling wave V_j^+ flowing into the network. All the other ports are terminated by a matched load. Since the ports are matched, the outward traveling waves are completely absorbed by the matched loads, and hence $V_k^+ = 0$ for $k \neq j$. With this arrangement, the outward traveling waves V_i^- are measured on all ports. Consequently, an entire column of the S-matrix can be computed as:

$$
S_{i,j} = \left. \frac{V_i^-}{V_j^+} \right|_{\substack{V_k^+=0 \\ k \neq j}}.
$$
(8.18)

The self term, $S_{j,j} = V_j^- / V_j^+$ represents the reflection coefficient of the j-th port with all other ports matched. From transmission line theory [1],

$$
S_{j,j} = \Gamma_j = \frac{Z_{in}^j - Z_0}{Z_{in}^j + Z_0}
$$
(8.19)

where Z_0 is the characteristic port impedance, and Z_{in}^j is the input impedance of port j with all other ports terminated in a matched load. Therefore, Z_{in}^j can be computed directly from $S_{j,j}$.

The magnitudes of the complex-valued scattering parameters can be related to power. That is,

$$
|S_{i,j}|^2 = \frac{|V_i^-|^2}{|V_j^+|^2} = \frac{|V_i^-|^2 / 2Z_0}{|V_j^+|^2 / 2Z_0} = \frac{P_i^-}{P_j^+}
$$
(8.20)

where P_i^- is the time average power flowing out of port i, and P_j^+ is the time average power flowing into port j. Therefore, the magnitude of $S_{i,j}$ represents the power flowing out of the i-th port relative to the power flowing into the j-th port.

For a lossless network, the power in the network must be conserved. Therefore, we must have that the net power flowing out of the network is equal to the power flowing in, or:

$$
\left. \sum_{i=1}^{N} |S_{i,j}|^2 \right|_{\text{lossless}} = 1.
$$
(8.21)

On the contrary, if the network is lossy, then the power lost within the network due to an excitation of the j−th port can be estimated as:

$$L_j = 1 - \sum_{i=1}^{N} \left|S_{i,j}\right|^2 \Bigg|_{\text{lossy}} . \tag{8.22}$$

The phase of $S_{i,j}$ represents the phase delay between the two ports. Therefore, the reference plane at which the inward or outward traveling waves are measured will impact the phase. When measuring the scattering parameters either physically or computationally with the FDTD method, it is not always possible to use the actual reference plane desired. Fortunately, the reference plane can be shifted. Refer to Fig. 8.4. Let x_i be the axial direction of each port and $x_i = 0$ be the coordinate of

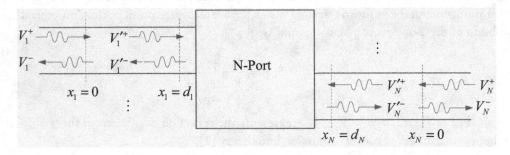

Figure 8.4: Shifting the reference planes of the N-port network.

the reference plane at which the scattering parameters were measured. Let $x_i = d_i$ be the coordinate of the new reference plane to which the S-parameters are to be shifted. The port is assumed to be a uniform transmission line. Therefore, one can let:

$$V_i'^{+} = e^{-j\beta_i d_i} V_i^{+} \tag{8.23}$$
$$V_i'^{-} = e^{+j\beta_i d_i} V_i^{-} \tag{8.24}$$

where $V_i'^{+}$ and $V_i'^{-}$ are the inward and outward traveling waves at the new reference plane, β_i is the propagation constant of the port waveguide, and a lossless line is assumed. This transformation can be applied to the scattering matrix (8.17), as:

$$
\begin{pmatrix}
e^{-j\beta_1 d_1} & 0 & \cdots & 0 \\
0 & e^{-j\beta_2 d_2} & \cdots & 0 \\
\vdots & \vdots & \ddots & \vdots \\
0 & 0 & \cdots & e^{-j\beta_N d_N}
\end{pmatrix}
\begin{pmatrix}
V_1'^{-} \\
V_2'^{-} \\
\vdots \\
V_N'^{-}
\end{pmatrix}
= [S]
\begin{pmatrix}
e^{j\beta_1 d_1} & 0 & \cdots & 0 \\
0 & e^{j\beta_2 d_2} & \cdots & 0 \\
\vdots & \vdots & \ddots & \vdots \\
0 & 0 & \cdots & e^{j\beta_N d_N}
\end{pmatrix}
\begin{pmatrix}
V_1'^{+} \\
V_2'^{+} \\
\vdots \\
V_N'^{+}
\end{pmatrix}
\tag{8.25}
$$

where $[S]$ is the scattering matrix. We can therefore transform the S-matrix as:

$$[S'] = \begin{pmatrix} e^{j\beta_1 d_1} & 0 & \cdots & 0 \\ 0 & e^{j\beta_2 d_2} & \cdots & 0 \\ \vdots & \vdots & \ddots & \vdots \\ 0 & 0 & \cdots & e^{j\beta_N d_N} \end{pmatrix} [S] \begin{pmatrix} e^{j\beta_1 d_1} & 0 & \cdots & 0 \\ 0 & e^{j\beta_2 d_2} & \cdots & 0 \\ \vdots & \vdots & \ddots & \vdots \\ 0 & 0 & \cdots & e^{j\beta_N d_N} \end{pmatrix}. \tag{8.26}$$

The scattering matrix transformed to the new reference planes, is expressed as:

$$\mathbf{V'}^- = [S'] \mathbf{V'}^+ \tag{8.27}$$

where $\mathbf{V'}^-$ and $\mathbf{V'}^+$ are the vectors of port inward and outward traveling waves.

Next, consider the case when the ports have different characteristic impedances. In this case, the scattering parameters are defined via normalized voltages, referred to as the wave amplitudes. To this end, let

$$a_i = V_i^+ / \sqrt{Z_{0i}}, \quad b_i = V_i^- / \sqrt{Z_{0i}} \tag{8.28}$$

represent the forward and backward traveling wave amplitudes, where Z_{0i} is the characteristic wave impedance of the i-th port. The total port voltage is represented via the complex wave amplitudes as:

$$V_i = V_i^+ + V_i^- = \sqrt{Z_{0i}} \, (a_i + b_i). \tag{8.29}$$

The generalized scattering matrix is introduced which represents the ratio of the inward and outward traveling wave amplitudes as:

$$\mathbf{a} = [S] \mathbf{b}. \tag{8.30}$$

A column of the scattering matrix is then again obtained by exciting the j-th port with a matched source (with impedance Z_{0j}), and terminating all other ports with a matched load Z_{0i}. The wave amplitudes on all ports are measured, and the j-th column of the scattering matrix is computed via:

$$S_{i,j} = \left. \frac{b_i}{a_j} \right|_{\substack{a_k=0 \\ k \neq j}}. \tag{8.31}$$

This can be expressed in terms of the port traveling wave voltages, as

$$S_{i,j} = \left. \frac{V_i^- \sqrt{Z_{0j}}}{V_j^+ \sqrt{Z_{0i}}} \right|_{\substack{V_k^+=0 \\ k \neq j}}. \tag{8.32}$$

8.2.3.1 Computing the S-Parameters via the FDTD Method

In order to directly compute the scattering parameters of a device using the FDTD method, one must be able to measure either the port voltages or currents, extract the port characteristic impedance, and terminate the ports with a matched load. Assuming the ports support a TEM or quasi-TEM mode, the port voltages and currents can be extracted from the time-dependent electromagnetic fields using the quasi-static approximaximations of (8.2) and (8.4), respectively. The frequency-dependent port voltages or currents can be evaluated via the FFT. The port impedance can also be computed by separately simulating a uniform line that is terminated with a matched load. The port characteristic impedance can then be computed via (8.8) or (8.12).

One way to terminate the ports with a matched load is to terminate the port with a resistive load, as detailed in Section 4.4.2. Unfortunately, at higher frequencies, this approach can be quite inaccurate. The actual load impedance will not be just the desired resistive load. Rather, there will be other stray reactances due to end effects as well as parasitic resistance due to radiation loss. Secondly, for quasi-TEM modes, the characteristic line impedance is frequency dependent. Consequently, the line will not be matched at all frequencies.

A more robust method is to actually terminate the port into an absorbing boundary. Most appropriate is the PML absorbing media, detailed in Chapter 6, since this provides minimum reflection error and can accurately terminate dispersive lines printed in a layered media. When terminating the port with the PML, it is effectively terminated with the true frequency-dependent characteristic port impedance. As an example, consider a port that is a microstrip line printed on a dielectric substrate. The microstrip line and the substrate will then extend all the way through the PML to the exterior terminating wall. In this way, the outward traveling wave is ideally matched by the PML media. If the port is an interior port that cannot be terminated by the PML, it should be terminated with a lumped resistance.

To measure the scattering parameters, all ports are terminated with a matched load (here assumed to be a PML). Then each column of the scattering matrix is measured by exciting one port at a time and measuring the port voltages in their reference planes due to the source excitation. For quasi-TEM lines, the port is typically excited with a soft voltage source (c.f., Section 4.4.1) since a soft source does not load the line. The source plane is located between the port reference plane and the terminating PML. Thus, the source will produce the forward traveling voltage wave flowing across the port reference plane.

The port voltage represents the outward traveling wave for all ports except for the source port. The source port voltage is a superposition of both inward and the outward traveling waves. In order to properly measure the scattering parameters, these must be separated. One way this can be accomplished is to perform an independent simulation of a uniform line with exactly the same dimensions as the port line that is terminated on both ends with a PML. The distance between the source plane and the port reference plane should be identical as compared to the case when simulating the device under test. This auxiliary simulation will only excite the inward traveling port voltage V_j^+ since the lines are matched via the PML. Subsequently, the outward traveling wave can be computed by subtracting the incident voltage from the total port voltage

$$V_j^- = V_j - V_j^+. \tag{8.33}$$

An alternate approach would be to extract the forward and backward propagating modes from the total voltage using an eigenvalue based method such as the matrix pencil method [5]. This approach is more efficient since it does not require an auxiliary simulation. However, it is subject to numerical error. The former approach is more accurate, but it does require the simulation of an auxiliary problem. Typically, the auxiliary problem is quite small compared to the full device simulation.

Once the time-dependent port voltages are computed, they are efficiently Fourier transformed using the FFT [2]. The frequency dependent scattering parameters are then computed via (8.18), or (8.32) if the port impedances are dissimilar.

The scattering parameters computed via this method are matched to the actual characteristic impedance Z_{0i} of each port. In some instances, it may be desired to transform the scattering parameters to a reference impedance that is different than Z_{0i}. For example, assume that 50 ohm port impedances are desired. However, due to dispersion the Z_{0i} are something other than 50 ohms. Fortunately, the scattering matrix computed via the FDTD method can be analytically transformed to a new scattering matrix referenced to the desired port impedances. To this end, let Z_{0i}^n be the new port impedances. Each port is extended with a zero length transmission line with characteristic impedance Z_{0i}^n that is terminated with a matched load impedance, Z_{0i}^n. The line extension imposes a reflection coefficient:

$$\Gamma_i = \frac{Z_{0i}^n - Z_{0i}}{Z_{0i}^n + Z_{0i}} \tag{8.34}$$

of the line voltages. With this configuration, the scattering parameters are computed analytically using (8.18) (or (8.32) if the Z_{0i}^n are different). That is, one port at a time is excited by a matched source (matched to Z_{0j}^n), and the outward traveling waves of all ports terminated via Z_{0i}^n are calculated. This is most easily done using signal flow diagrams [1]. The result for an arbitrary N-port network is summarized here. Let $[\Gamma]$ be the reflection coefficient matrix, where

$$[\Gamma] = \begin{bmatrix} \Gamma_1 & 0 & 0 & 0 \\ 0 & \Gamma_2 & 0 & 0 \\ \vdots & \vdots & \ddots & \vdots \\ 0 & 0 & \cdots & \Gamma_N \end{bmatrix} \tag{8.35}$$

and Γ_i is the reflection coefficient of the i−th port given by (8.34). Next, let $[A]$ be the amplitude matrix, defined as:

$$[A] = \begin{bmatrix} A_1 & 0 & 0 & 0 \\ 0 & A_2 & 0 & 0 \\ \vdots & \vdots & \ddots & \vdots \\ 0 & 0 & \cdots & A_N \end{bmatrix} \tag{8.36}$$

where

$$A_i = \sqrt{\frac{Z_{0i}^n}{Z_{0i}}} \frac{1}{Z_{0i}^n + Z_{0i}}. \tag{8.37}$$

The scattering matrix transformed to the new port impedances is then found to be:

$$[S^n] = [A]^{-1} ([S] - [\Gamma]) ([I] - [\Gamma][S])^{-1} [A] \qquad (8.38)$$

where $[I]$ is the identity matrix, and $[S^n]$ is the transformed scattering matrix.

8.2.3.2 Example S-Parameter Calculation Via the FDTD Method

As an example, the FDTD method is used to compute the scattering parameters of the 3 dB Wilkinson power divider [1] illustrated in Fig. 8.5. This is a three-port microstrip device with

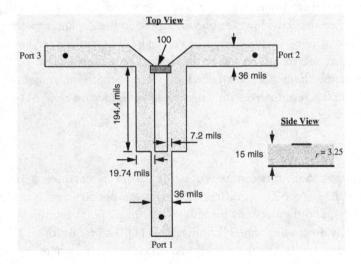

Figure 8.5: Microstrip Wilkinson power divider printed on a dielectric substrate.

50 ohm ports. The device was designed for 9 GHz operation. Thus, at 9 GHz, the divider should equally divide power entering port 1 between ports 2 and 3. The outputs are also in phase. Ideally, at 9 GHz ports 1, 2 and 3 are matched (i.e., $|S_{i,i}| = 0$), and ports 2 and 3 are isolated (i.e., $|S_{2,3}| = 0$).

The FDTD simulation was performed with a non-uniform grid to better match the geometry. The 100 ohm isolation resistance was modeled as a lumped resistance in the FDTD model. The lines were matched with a PML absorber. For each port excitation, the incident wave was computed using a separate simulation of a uniform line.

The scattering parameters versus frequency computed via the FDTD method are illustrated in Fig. 8.6. Similarly, Fig. 8.7 shows a comparison result computed using the Agilent Advanced Design Software® when the device is modeled via empirical transmission line models. The two results compare relatively well. However, the FDTD model will be more accurate since it is based on a true full wave model of the device.

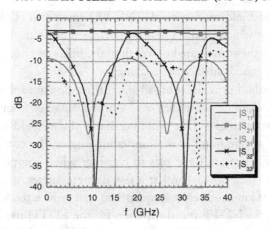

Figure 8.6: Magnitude of the scattering parameters of the Wilkinson power-divider computed via the FDTD method.

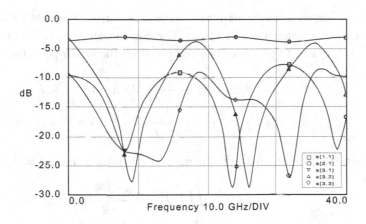

Figure 8.7: Magnitude of the scattering parameters of the Wilkinson power-divider computed via Agilent Advanced Design System® using an empirical model.

8.3   NEAR-FIELD TO FAR-FIELD (NF-FF) TRANSFORMATIONS

Often, it is necessary to compute the far-field pattern of a radiator or scatterer. However, the FDTD method only provides a solution for the electromagnetic fields in the near vicinity of the device under test. Since it is not practical to extend the computational domain into the far field, a manner by which one can predict the far fields using only near fields must be developed. This can efficiently be done using a "Near-Field to Far-Field" (NF-FF) transform. This transformation is based on Green's second identity, which predicts that if the tangential fields on a closed surface radiated by a set of sources bound by the surface are known, then the fields exterior to that surface can be uniquely predicted from the tangential surface fields alone.

The FDTD method provides a broad-band time-domain solution for the near fields. Consequently, the far fields can be computed over a broad range of frequencies. The far field can also be computed at any set of angles. One will find that if the far field is computed at a large number of frequencies and at a large number of observation angles, the computational effort can become quite large. Therefore, there are trade offs to consider when computing the far field. Knowing *a priori* which is needed (i.e., broad-band at a few discrete angles, or a large number of angles at a moderately small set of known discrete frequencies), can reduce the level of computational effort.

In this section, the NF-FF transformation for the FDTD method is explored. The foundational theory is initially introduced in Section 8.3.1. Efficient NF-FF transforms for the FDTD method in the frequency domain are detailed in Section 8.3.2.

8.3.1   HUYGEN SURFACE

Consider a set of impressed, time-harmonic current densities $\vec{J}^i$ and $\vec{M}^i$, radiating in a homogeneous free space with material profile (μ, ε). The current densities radiate the fields $\vec{E}, \vec{H}$ in the homogeneous free space. The fields can be computed via the vector potentials [8] as:

$$\vec{E}(\vec{r}) = -j\omega\mu\vec{A}(\vec{r}) + \frac{1}{j\omega\varepsilon}\nabla\nabla\cdot\vec{A}(\vec{r}) - \nabla\times\vec{F}(\vec{r}), \tag{8.39}$$

$$\vec{H}(\vec{r}) = -j\omega\varepsilon\vec{F}(\vec{r}) + \frac{1}{j\omega\mu}\nabla\nabla\cdot\vec{F}(\vec{r}) + \nabla\times\vec{A}(\vec{r}). \tag{8.40}$$

$\vec{A}$ is the magnetic vector potential and $\vec{F}$ is the electric vector potential, which are defined as:

$$\vec{A}(\vec{r}) = \iiint\limits_V \vec{J}^i(\vec{r}')\frac{e^{-jkR}}{4\pi R}dv', \tag{8.41}$$

$$\vec{F}(\vec{r}) = \iiint\limits_V \vec{M}^i(\vec{r}')\frac{e^{-jkR}}{4\pi R}dv', \tag{8.42}$$

where $\vec{r}'$ is the source coordinate and $\vec{r}$ is the observation coordinate, $R = |\vec{r} - \vec{r}'|$ is the distance between the two, and $k = \omega\sqrt{\mu\varepsilon}$ is the wave number.

Now, let S be a fictitious surface that completely envelopes the sources $\vec{J}^i$ and $\vec{M}^i$, as illustrated in Fig. 8.8, and let $\vec{r}_s$ be a point on S. Then, let

$$\vec{J}_s(\vec{r}_s) = \hat{n}\times\vec{H}(\vec{r}_s), \tag{8.43}$$

and

$$\vec{M}_s(\vec{r}_s) = \vec{E}(\vec{r}_s)\times\hat{n}, \tag{8.44}$$

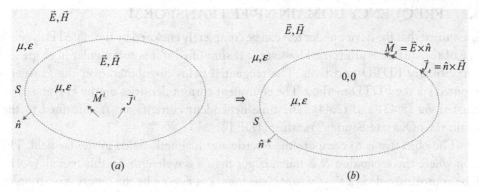

Figure 8.8: Equivalence principal. (a) Original problem with impressed currents radiating in a homogenous free space. (b) Equivalent currents distributed on the Huygen surface S radiating in a homogeneous free space.

be *equivalent current densities* [8] on S, where $\hat{n}$ is the unit normal directed out of the volume bound by S, and $\vec{E}$ and $\vec{H}$ are fields radiated by the impressed current densities. The surface S is referred to as a Huygen surface. Green's second identity predicts that if allowed to radiate in the homogeneous free space (μ, ε), $\vec{J}_s$ and $\vec{M}_s$ will exactly radiate $\vec{E}$, $\vec{H}$ in the region outside of S and null fields inside of S [9]. The fields radiated by the surface current densities are computed by (8.39) and (8.40), with vector potentials arising from surface current densities:

$$\vec{A}\left(\vec{r}\right) = \iint_S \vec{J}_s\left(\vec{r}'\right) \frac{e^{-jkR}}{4\pi R} ds', \tag{8.45}$$

$$\vec{F}\left(\vec{r}\right) = \iint_S \vec{M}_s\left(\vec{r}'\right) \frac{e^{-jkR}}{4\pi R} ds', \tag{8.46}$$

where $\vec{r}'$ is the source point on S.

Thus, if the tangential fields are known over the entire Huygen surface, then these "equivalent currents" can be used to predict the field outside of the surface S (illustrated in Fig. 8.8). This principal is known as the surface equivalence principal [8]. The surface equivalence principal is applied to the FDTD method to compute the far-fields. That is, a Huygen surface is introduced that surrounds the device under test. The FDTD is then used to compute the fields on the Huygen surface, providing the equivalent current densities. The current densities are then radiated exterior to the Huygen surface, presumably into the far field which extends beyond the FDTD computational domain.

8.3.2   FREQUENCY DOMAIN NF-FF TRANSFORM

It is assumed that the device under test can be completely enclosed by the closed Huygen surface. The shape of the surface is arbitrary. However, it is simplified if it is rectangular in shape and conforms to the primary FDTD grid faces. The tangential fields distributed over the Huygen surface are computed via the FDTD method. The equivalent current densities on the Huygen surface are then evaluated via (8.43) and (8.44). The time-dependent currents are transformed to the frequency domain via a Discrete Fourier Transform (DFT).

The objective is to compute the electric and magnetic fields in the far field. That is, in the region where the separation R is much larger than a wavelength. In this region, (8.41) and (8.42) can be approximated via a *far-field approximation*. Let $\vec{r} = r\hat{r}$ be the observation coordinate relative to a global origin, where r is the radial distance from the origin, and $\hat{r}$ is the unit radial spherical vector. Similarly, let $\vec{r}' = r'\hat{r}'$ be the source coordinate relative to the same origin. Then,

$$R = \left| \vec{r} - \vec{r}' \right| = \left[r^2 + r'^2 - 2r\,r' \left(\hat{r} \cdot \hat{r}' \right) \right]^{\frac{1}{2}}, \tag{8.47}$$

where

$$\hat{r} \cdot \hat{r}' = \cos\psi, \tag{8.48}$$

and ψ is the solid angle between the source and observation vectors, as illustrated in Fig. 8.9. In the

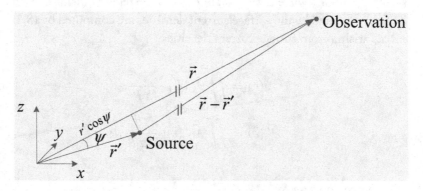

Figure 8.9: Far field approximation of the separation vector.

region where $r >> r'$, we can approximate $R \approx r$. This is sufficient for measurements of magnitude. However, phase is much more sensitive to small changes. Thus, when measuring phase, it is more accurate to approximate $R \approx r - r' \cos\psi$. Therefore, from (8.45) or (8.46), in the far field, the approximation is made where [8]:

$$\left. \frac{e^{-jkR}}{4\pi R} \right|_{\text{Farfield}} \approx \frac{e^{-jkr}e^{jkr'(\hat{r}\cdot\hat{r}')}}{4\pi r}, \tag{8.49}$$

The electric and magnetic fields are computed using (8.39) and (8.40) with the far-field approximation (8.49) used within (8.45) and (8.46). This leads to the far field expressions [8]:

$$E_r \approx 0, \tag{8.50}$$

$$E_\theta \approx -jk\frac{e^{-jkr}}{4\pi r}\left(L_\phi + \eta N_\theta\right), \tag{8.51}$$

$$E_\phi \approx jk\frac{e^{-jkr}}{4\pi r}\left(L_\theta - \eta N_\phi\right), \tag{8.52}$$

$$H_r \approx 0, \tag{8.53}$$

$$H_\theta \approx jk\frac{e^{-jkr}}{4\pi r}\left(N_\phi - \frac{1}{\eta}L_\theta\right) = -\frac{1}{\eta}E_\phi, \tag{8.54}$$

$$H_\phi \approx -jk\frac{e^{-jkr}}{4\pi r}\left(N_\theta + \frac{1}{\eta}L_\phi\right) = \frac{1}{\eta}E_\theta, \tag{8.55}$$

where, the fields are expressed in a spherical coordinate system and

$$\vec{N}(\theta, \phi) \approx \int_S \vec{J}_s\left(f, \vec{r}'\right) e^{j\vec{k}\cdot\vec{r}'}ds', \tag{8.56}$$

$$\vec{L}(\theta, \phi) \approx \int_S \vec{M}_s\left(f, \vec{r}'\right) e^{j\vec{k}\cdot\vec{r}'}ds', \tag{8.57}$$

where, $\vec{r}'$ is the position vector of the source coordinate and

$$\vec{k} = k\hat{r} = k\left(\sin\theta\cos\phi\hat{x} + \sin\theta\sin\phi\hat{y} + \cos\theta\hat{z}\right) \tag{8.58}$$

is referred to as the wave vector (or k-vector). Note that (θ, ϕ) describes the angular location of the observation point $\vec{r}$ in spherical coordinates, where θ is the zenith angle off the z-axis, and ϕ is the azimuthal angle between the x-axis and the projection of $\vec{r}$ in the $x - y$ plane. The fields are projected onto the spherical vectors, where $\hat{r}$ is defined in (8.58), and

$$\hat{\theta} = \cos\theta\cos\phi\hat{x} + \cos\theta\sin\phi\hat{y} - \sin\theta\hat{z}, \tag{8.59}$$

and,

$$\hat{\phi} = -\sin\phi\hat{x} + \cos\phi\hat{y}. \tag{8.60}$$

In this approximation, only fields with a $1/r$ dependence are maintained. Fields with $1/r^2$ dependencies or greater are neglected, since it is assumed in the far-field their amplitudes are negligibly small. As a consequence, the radial-directed fields are negligibly small. Since the far-fields are traveling in the radial direction, they are essentially spherical plane waves. As observed in (8.54) and (8.55), the ratio of orthogonal components of E and H is the characteristic wave impedance. Thus, in general, only E need be computed in the far field. The magnetic field can be computed from E using a scaling by the wave impedance.

The surface current densities are obtained from the near magnetic and electric fields in the discrete FDTD space. The principal Huygen surface is defined by the primary grid. Assuming the grid is rectangular in dimension, it is bound by coordinates (i_{s1}, j_{s1}, k_{s1}) and (i_{s2}, j_{s2}, k_{s2}).

An example of this is illustrated in Fig. 8.9. The current densities are defined relative to this surface. For example, on the shaded surface of Fig. 8.9, where $\hat{n} = \hat{z}$,

$$\hat{y} M_y^{n+\frac{1}{2}}\bigg|_{i+\frac{1}{2},j,k_{s2}} = \hat{x} E_x^{n+\frac{1}{2}}\bigg|_{i+\frac{1}{2},j,k_{s2}} \times \hat{z} = -\hat{y} E_x^{n+\frac{1}{2}}\bigg|_{i+\frac{1}{2},j,k_{s2}}, \tag{8.61}$$

and

$$\hat{x} M_x^{n+\frac{1}{2}}\bigg|_{i,j+\frac{1}{2},k_{s2}} = \hat{y} E_y^{n+\frac{1}{2}}\bigg|_{i,j+\frac{1}{2},k_{s2}} \times \hat{z} = \hat{x} E_y^{n+\frac{1}{2}}\bigg|_{i,j+\frac{1}{2},k_{s2}}, \tag{8.62}$$

where k_s is the discrete coordinate of S. Similar expressions can be derived for the other surfaces.

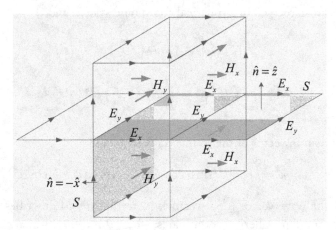

Figure 8.10: Huygen surface defined by the primary FDTD grid.

The far-field vector potential $\vec{L}(\theta, \phi)$ in (8.57) requires the integration over the Huygen surface. This is done numerically by assuming the integrand to be constant over a grid cell. The integration can be performed by superimposing integrations over each of the six faces of S. For example, the partial contribution to L_x from the top face in the k_{s2} plane is approximated as:

$$L_x(\theta, \phi)|_{k_{s2}} \approx \sum_{j=j_{s1}}^{j_{s2}-1} \sum_{i=i_{s1}}^{i_{s2}} v_{i_{s1},i_{s2}}^i E_y\bigg|_{i,j+\frac{1}{2},k_{s2}} e^{j\vec{k}\cdot\vec{r}'_{i,j+\frac{1}{2},k_{s2}}} \Delta x \Delta y \tag{8.63}$$

where

$$v_{i_{s1},i_{s2}}^i = \begin{cases} \frac{1}{2}, & \text{if } i = i_{s1}, \text{ or } i = i_{s2} \\ 1, & \text{else} \end{cases} \tag{8.64}$$

and

$$\vec{r}'_{i,j+\frac{1}{2},k_{s2}} = i\Delta x\hat{x} + (j + \tfrac{1}{2})\Delta y\hat{y} + k_{s2}\Delta z\hat{z} \tag{8.65}$$

is the position vector of the source coordinate at the edge centers, and the k-vector, defined by (8.58), is a function of the observation angle. Note that $v^i_{i_{s1},i_{s2}}$ accounts for the edge terms where E_y is at a corner and only one-half of the discrete face is on the Huygen surface. This is illustrated in Fig. 8.9. Similar expressions can also be derived for the remaining contributions from the other five surfaces, noting that there is no contribution to L_x from the two x-normal surfaces. Similar expressions can be derived for L_y and L_z.

The evaluation of $\widetilde{N}(\theta, \phi)$ is a little more difficult. What is realized is that the magnetic fields are one-half of a cell removed from the Huygen surface due to the spatial staggering of the grid. Consequently, an approximation must be made. What is typically done is to form two surfaces – one-half of a cell above the Huygen surface and the other one-half a cell below. An electric current density is then expressed on both surfaces. If we approximate that the local magnetic field is a plane wave traveling outward across the boundary, then the electric current densities on the surfaces above and below will be of the form:

$$J_{i,j,k_{s2}+\frac{1}{2}} = Je^{-j\beta_x i\Delta x - j\beta_y j\Delta y - j\beta_z(k_{s2}+\frac{1}{2})\Delta z}, \quad J_{i,j,k_{s2}-\frac{1}{2}} = Je^{-j\beta_x i\Delta x - j\beta_y j\Delta y - j\beta_z(k_{s2}-\frac{1}{2})\Delta z}. \tag{8.66}$$

Consequently, if the current on the Huygen surface is approximated via the geometric mean, then:

$$J_{i,j,k_{s2}} = \sqrt{J_{i,j,k_{s2}+\frac{1}{2}} J_{i,j,k_{s2}-\frac{1}{2}}} = Je^{-j\beta_x i\Delta x - j\beta_y j\Delta y - j\beta_z k_{s2}\Delta z} \tag{8.67}$$

which is exactly what is desired. Thus, if the electric currents are approximated via the geometric mean, the error due to grid staggering can be reduced [10]. Consequently, the partial contribution to the electric far-field vector potential from the top face is expressed as:

$$N_x(\theta, \phi)|_{k_{s2}} \approx -\sum_{j=j_{s1}}^{j_{s2}} \sum_{i=i_{s1}}^{i_{s2}-1} v^j_{j_{s1},j_{s2}} \sqrt{H_{y_{i+\frac{1}{2},j,k_{s2}+\frac{1}{2}}} H_{y_{i+\frac{1}{2},j,k_{s2}-\frac{1}{2}}}} e^{j\vec{k}\cdot\vec{r}'_{i+\frac{1}{2},j,k_{s2}}} \Delta x \Delta y. \tag{8.68}$$

A less accurate approximation that can also be used, is a linear average of the two surface current densities. In this case,

$$N_x(\theta, \phi)|_{k_{s2}} \approx -\frac{1}{2}\sum_{j=j_{s1}}^{j_{s2}} \sum_{i=i_{s1}}^{i_{s2}-1} v^j_{j_{s1},j_{s2}} \left(H_{y_{i+\frac{1}{2},j,k_{s2}+\frac{1}{2}}} + H_{y_{i+\frac{1}{2},j,k_{s2}-\frac{1}{2}}}\right) e^{j\vec{k}\cdot\vec{r}'_{i+\frac{1}{2},j,k_{s2}}} \Delta x \Delta y. \tag{8.69}$$

Similar expressions can also be derived for the remaining contributions from the other five surfaces, noting that there is no contribution from the x-normal surfaces. Similar expressions can be derived for N_y and N_z.

An important consideration in the NF-FF transform is how to efficiently Fourier transform the tangential fields on the Huygen surface. If a broad frequency response is desired, one can compute $\vec{J}_s\,(\vec{r}_s, t)$ and $\vec{M}_s\,(\vec{r}_s, t)$ at all positions on the Huygen surface for all time, and then use the FFT to transform the currents to $\vec{J}_s\,(\vec{r}_s, \omega)$ and $\vec{M}_s\,(\vec{r}_s, \omega)$. For very large problems, where the size of the surface mesh is large, and the number of time-steps is large, this requires a lot of data. Typically, $\vec{J}_s\,(\vec{r}_s, t)$ and $\vec{M}_s\,(\vec{r}_s, t)$ will have to be written to an external file first, and then Fourier transformed after the completion of the FDTD simulation. The Fourier transforms of the tangential near fields are most efficiently performed via the FFT. Then, at desired frequencies, the far field can be computed using the prescribed algorithm at as many observation angles as desired. In this case, (8.68) should be used to most accurately compute the far electric vector potential.

On the other hand, if the far field is desired at a finite number of discrete frequencies and if the set of observation angles is also known *a priori*, then the far field can be computed via a Discrete Fourier Transform (DFT). To this end, let $\vec{M}\,(t, \vec{r}')$ represent the time-dependent electric current density. The frequency domain current can then be approximated via the Fourier transform as:

$$\vec{M}\,(f_i, \vec{r}') = \int_0^\infty \vec{M}\,(t, \vec{r}')\, e^{-j2\pi f_i t}\, dt \tag{8.70}$$

where, f_i is the discrete frequency at which the current is being evaluated. The integration can be approximated via a discrete integration method:

$$\vec{M}\,(f_i, \vec{r}') \approx \Delta t \sum_{n=0}^N \vec{M}\left(\left(n + \tfrac{1}{2}\right)\Delta t, \vec{r}'\right) e^{-j2\pi f_i\left(n+\frac{1}{2}\right)\Delta t}. \tag{8.71}$$

Combining this with (8.63)

$$L_x\,(\theta, \phi) \approx \Delta t \sum_{n=0}^N e^{-j2\pi f_i\left(n+\frac{1}{2}\right)\Delta t} \left(\begin{array}{l} \Delta x\,\Delta y \displaystyle\sum_{j=j_{s1}}^{j_{s2}-1} \sum_{i=i_{s1}}^{i_{s2}} v^i_{i_{s1},i_{s2}} E_y{}^{n+\frac{1}{2}}_{i,j+\frac{1}{2},k_{s2}} e^{j\vec{k}\cdot\vec{r}'_{i,j+\frac{1}{2},k_{s2}}} - \\[2mm] \Delta x\,\Delta y \displaystyle\sum_{j=j_{s1}}^{j_{s2}-1} \sum_{i=i_{s1}}^{i_{s2}} v^i_{i_{s1},i_{s2}} E_y{}^{n+\frac{1}{2}}_{i,j+\frac{1}{2},k_{s1}} e^{j\vec{k}\cdot\vec{r}'_{i,j+\frac{1}{2},k_{s1}}} + \\[2mm] \Delta x\,\Delta z \displaystyle\sum_{k=k_{s1}}^{k_{s2}-1} \sum_{i=i_{s1}}^{i_{s2}} v^i_{i_{s1},i_{s2}} E_z{}^{n+\frac{1}{2}}_{i,j_{s1},k+\frac{1}{2}} e^{j\vec{k}\cdot\vec{r}'_{i,j_{s1},k+\frac{1}{2}}} - \\[2mm] \Delta x\,\Delta z \displaystyle\sum_{k=k_{s1}}^{k_{s2}-1} \sum_{i=i_{s1}}^{i_{s2}} v^i_{i_{s1},i_{s2}} E_z{}^{n+\frac{1}{2}}_{i,j_{s2},k+\frac{1}{2}} e^{j\vec{k}\cdot\vec{r}'_{i,j_{s2},k+\frac{1}{2}}} \end{array} \right) \tag{8.72}$$

where this is computed for each desired angle pair (θ, ϕ). Similar expressions can be derived for L_y and L_z. Consequently, $\vec{L}\,(\theta, \phi)$ can be updated at each time-step for each frequency f_i. To reduce storage of $\vec{L}\,(\theta, \phi)$, only the two $\hat{\theta}$ and $\hat{\phi}$-projections need be stored.

The DFT of the electric current density can similarly be computed. To this end:

$$\vec{J}\left(f_i, \vec{r}'\right) \approx \Delta t \sum_{n=0}^{N} v_0^n \vec{J}\left(n\Delta t, \vec{r}'\right) e^{-j2\pi f_i n \Delta t} \tag{8.73}$$

where $v_0^n = \frac{1}{2}$ if $n = 0$, else it is $= 1$. This can also be combined with the discrete surface integral over the Huygen surface. It was noted that in order to account for the staggering of the fields, the electric current density would have to be interpolated using field values above and below the Huygen surface. While the geometric mean (c.f., (8.68)) is most accurate, it is a non-linear expression. Rather, the linear average method of (8.69) will have to be employed. Consequently, the discrete time integration and surface integrations can be swapped, and an expression similar to (8.72) can be derived to compute $\vec{N}(\theta, \phi)$ at a discrete set of angles.

Once the simulation has reached a steady-state, and L_θ, L_ϕ, N_θ, and N_ϕ are converged, then the far fields can be computed over the discrete range of angles via (8.50)–(8.55).

8.3.3 ANTENNA GAIN

The antenna gain is the ratio of the power density of an antenna radiated in a given direction to the power density of an isotropic radiator that uniformly radiates the same total power. To this end, let P_{in} be the input power of an antenna. For example, consider an antenna driven by a single port. Let $S_{1,1}$ be scattering parameter of the 1-port network. Then,

$$P_{in} = (1 - |S_{1,1}|^2) \frac{|V_1^+|^2}{2Z_{01}}. \tag{8.74}$$

Next, consider an isotropic radiator that radiates the power uniformly over all angles. The power density at any angle (θ, ϕ) due to this isotropic radiator has a power density of:

$$P^{iso} = \frac{P_{in}}{4\pi r^2}. \tag{8.75}$$

where, r, is the radial distance from the origin.

The actual antenna radiates power with an angular dependence. As discussed in the previous section, the far field of the antenna can be computed using the NF-FF transform. The power density at any point can be computed as:

$$P^{ant}(r, \theta, \phi) = \frac{1}{2}\text{Re}\left\{\vec{E}(r, \theta, \phi) \times \vec{H}(r, \theta, \phi) \cdot \hat{r}(\theta, \phi)\right\}. \tag{8.76}$$

In the far field, this is approximated as:

$$P^{ant}(r, \theta, \phi) = P_\theta^{ant}(r, \theta, \phi) + P_\phi^{ant}(r, \theta, \phi), \tag{8.77}$$

where

$$P_\theta^{ant}(r, \theta, \phi) = \frac{|E_\theta(r, \theta, \phi)|^2}{2\eta_0}, \tag{8.78}$$

$$P_\phi^{ant}(r, \theta, \phi) = \frac{|E_\phi(r, \theta, \phi)|^2}{2\eta_0}, \tag{8.79}$$

and E_θ and E_ϕ are defined by (8.51) and (8.52).

Finally, the antenna gain is defined as:

$$G_\theta^{ant}(\theta, \phi) = \frac{P_\theta^{ant}(r, \theta, \phi)}{P^{iso}(r, \theta, \phi)} = \frac{4\pi r^2}{P_{in}} \frac{|E_\theta(r, \theta, \phi)|^2}{2\eta_0} = \frac{k^2}{4\pi} \frac{|L_\phi(\theta, \phi) + \eta_0 N_\theta(\theta, \phi)|^2}{2\eta_0 P_{in}}, \tag{8.80}$$

and

$$G_\phi^{ant}(\theta, \phi) = \frac{P_\phi^{ant}(r, \theta, \phi)}{P^{iso}(r, \theta, \phi)} = \frac{4\pi r^2}{P_{in}} \frac{|E_\phi(r, \theta, \phi)|^2}{2\eta_0} = \frac{k^2}{4\pi} \frac{|L_\theta(\theta, \phi) - \eta_0 N_\phi(\theta, \phi)|^2}{2\eta_0 P_{in}}. \tag{8.81}$$

In this form, the antenna gain is decoupled into two polarizations. The total antenna gain would be the superposition of the polarized gains:

$$G^{ant}(\theta, \phi) = G_\theta^{ant}(\theta, \phi) + G_\phi^{ant}(\theta, \phi). \tag{8.82}$$

8.3.4   SCATTERING CROSS SECTION

The scattering cross section of a target is an equivalent area that reflects the transmitted power in a given direction. The scattering cross section is typically computed by illuminating the target with a uniform plane wave, and then computing the field scattered by the target in a specific direction. If the scattering direction is back towards the transmitter of the incident plane wave, it is referred to as the monostatic cross section. If the scattering direction is a set of arbitrary angles, it is referred to as a bistatic cross section.

The scattering cross section is computed in the following manner. Consider a vertically polarized incident plane wave (that is, $\vec{E}_V^{inc} = \hat{\theta} E_\theta^{inc}(\theta^{inc}, \phi^{inc})$). Let P_θ^{inc} be the power density of the incident plane wave:

$$P_\theta^{inc} = \frac{|E_\theta^{inc}(\theta^{inc}, \phi^{inc})|^2}{2\eta_0}. \tag{8.83}$$

Let, $P_\theta^{scat}(r, \theta, \phi)$ be the power density of the vertically polarized scattered field evaluated in the far-field at (r, θ, ϕ) due to the θ-directed electric field. That is:

$$P_\theta^{scat}(r, \theta, \phi) = \frac{|E_\theta^{scat}(r, \theta, \phi)|^2}{2\eta_0}. \tag{8.84}$$

The scattering cross section is expressed as the ratio of the scattered power density to the incident power density times the area of a sphere of radius r, in the limit $r \to \infty$ [11]. That is,

$$\sigma_{\theta,\theta}(\theta,\phi) = \lim_{r\to\infty} 4\pi r^2 \frac{P_\theta^{scat}(r,\theta,\phi)}{P_\theta^{inc}} = \lim_{r\to\infty} 4\pi r^2 \frac{\left|E_\theta^{scat}(r,\theta,\phi)\right|^2}{2\eta_0}\frac{1}{P_\theta^{inc}}. \tag{8.85}$$

Expanding E_θ^{scat} using the far field approximation of (8.51) along with (8.83), this is more appropriately expressed as

$$\sigma_{\theta,\theta}(\theta,\phi) = \frac{k^2}{4\pi}\frac{\left|L_\phi(\theta,\phi)+\eta_0 N_\theta(\theta,\phi)\right|^2}{\left|E_\theta^{inc}(\theta^{inc},\phi^{inc})\right|^2}. \tag{8.86}$$

where $L_\phi(\theta,\phi)$ and $N_\theta(\theta,\phi)$ are computed using the NF-FF transform.

Similarly, let $P_\phi^{scat}(r,\theta,\phi)$ be the power density of the scattered field evaluated at (r,θ,ϕ) due to the ϕ-directed electric field:

$$P_\phi^{scat}(r,\theta,\phi) = \frac{\left|E_\phi^{scat}(r,\theta,\phi)\right|^2}{2\eta_0}. \tag{8.87}$$

The phi-polarized (or horizontal) scattering cross section due to the vertically polarized incident wave is expressed as:

$$\sigma_{\phi,\theta}(\theta,\phi) = \frac{k^2}{4\pi}\frac{\left|L_\theta(\theta,\phi)-\eta_0 N_\phi(\theta,\phi)\right|^2}{\left|E_\theta^{inc}(\theta^{inc},\phi^{inc})\right|^2}. \tag{8.88}$$

The scattering cross section can also be computed for a horizontally polarized incident plane wave (that is, $\vec{E}_H^{inc} = \hat{\phi}E_\phi^{inc}(\theta^{inc},\phi^{inc})$). The scattering cross-sections are then defined as:

$$\sigma_{\theta,\phi}(\theta,\phi) = \frac{k^2}{4\pi}\frac{\left|L_\phi(\theta,\phi)+\eta_0 N_\theta(\theta,\phi)\right|^2}{\left|E_\phi^{inc}(\theta^{inc},\phi^{inc})\right|^2}, \tag{8.89}$$

and

$$\sigma_{\phi,\phi}(\theta,\phi) = \frac{k^2}{4\pi}\frac{\left|L_\theta(\theta,\phi)-\eta_0 N_\phi(\theta,\phi)\right|^2}{\left|E_\phi^{inc}(\theta^{inc},\phi^{inc})\right|^2}, \tag{8.90}$$

where $L_\phi(\theta,\phi)$ and $N_\theta(\theta,\phi)$ are computed using the NF-FF transform due to the horizontally polarized incident wave.

8.4   PROBLEMS

1. You are to extract the characteristic impedance of a microstrip line using the FDTD method. The microstrip line is printed on a conductor backed 10 mil[1] thick alumina substrate ($\varepsilon_r = 9.8$).

[1] 1 mil = 0.001 inches.

The substrate is assumed to be lossless. The microstrip line has a width of 9.7 mils, and is assumed to be infinitesimally thin. The uniform line is 100 mils long and should be matched on both ends using a perfectly matched layer (c.f., Chapter 6). (Note that the uniform line should extend completely through the PML to the terminating walls.) Excite the line using a soft voltage source 10 mils from one of the PML boundaries (Chapter 4). The source should have a Gaussian-pulse signature with a 10 GHz bandwidth. Compute the line voltage and current roughly 10 mils from the far PML boundary. The length of the simulation should be sufficiently long that the line voltage and current reach a steady state value near zero. Use the FFT to Fourier transform the line voltage and current. (In order to avoid aliasing, the time-dependent signal should be "zero-padded" such that the total length of the FFT is at least twice the length of the time-dependent signal.) If Δt is the time-step, and the FFT is of length N_{FFT}, then the discrete frequency dependent signal resulting from the FFT will be uniformly sampled with spacing $\Delta f = 1/(N_{FFT} \Delta t)$. Use two different discretizations (e.g., 2 space cells across the width and through the substrate, and 6 space cells across the width and through the substrate). Plot the impedance versus frequency over the range 0 to 10 GHz using both discretizations. Compare the line impedance when computed using (8.5) or (8.8). (The line impedance should be approximately 49.96 Ω.) Do you expect frequency dependence in the line impedance? Why?

2. Using the microstrip line from problem 1, you are to extract the effective propagation constant of the quasi-TEM mode. To do this, compute the line voltage at two discrete points on the line separated by 60 mils. Let the point closest to the source be V_1, and the other line voltage be designated as V_2. Since the line is matched via the PML, only a forward traveling wave should be traveling down the line. Thus, we expect that:

$$\frac{V_2(\omega)}{V_1(\omega)} = e^{-\alpha d} e^{-j\beta d}$$

where d = 60 mils, α is the attenuation constant (Np/m) and β is the wave number (rad/m). Plot β versus frequency over the range of 0 to 10 GHz. From β, extract the effective phase velocity of the quasi-TEM wave.

3. Derive the transformation of the scattering matrix port impedance in (8.38) for a general N-port network.

4. Using the FDTD method, compute the scattering parameters of the Wilkinson power divider in Figure 8.5 (i.e., reproduce the results in Figure 8.6). Assume that the microstrip and ground plane are perfectly conducting. Also, assume that the microstrip is infinitesimally thin. Use a lumped resistive load to model the 100 Ω isolation impedance. Assume 50 ohm port impedances.

5. Starting with (8.39), (8.40), (8.45), and (8.46), and the far-field approximation (8.49), derive (8.50)–(8.57).

6. Derive the expressions for the y and z projections of the far-field vector potential $L_y(\theta,\varphi)$ and $L_z(\theta,\varphi)$ computed in the discrete FDTD space similar to (8.63), but taking into account all 6 sides of the discrete Huygen surface.

7. Derive the expressions for the far-field vector potentials $N_x(\theta,\varphi)$, $N_y(\theta,\varphi)$, and $N_z(\theta,\varphi)$ computed in the discrete FDTD space using the geometric mean expression similar to (8.68). Take into account all 6 sides of the discrete Huygen surface.

8. Using (8.51), (8.83)–(8.85), derive (8.86). Similarly, using (8.52), derive (8.88).

REFERENCES

[1] D. M. Pozar, *Microwvae Engineering*. NJ: John Wiley & Sons, Inc., 2005. 173, 178, 179, 183, 184

[2] W. H. Press, B. P. Flannery, S. A. Teukolsky, and W. T. Vetterling, *Numerical Recipe's: The Art of Scientific Computing*, 2nd ed. New York: Cambridge University Press, 1992. 175, 183

[3] J. Y. Fang and D. W. Xeu, "Numerical errors in the computation of impedances by FDTD methods and ways to eliminate them," *IEEE Microwave and Guided Wave Letters*, vol. 5, pp. 6–8, Jan 1995. DOI: 10.1109/75.382377 176

[4] Z. Altman, R. Mittra, O. Hashimoto, and E. Michielssen, "Efficient representation of induced currents on large scatterers using the generalized pencil of function method," *IEEE Transactions on Antennas and Propagation*, vol. 44, pp. 51–57, Jan 1996. DOI: 10.1109/8.477528 178

[5] T. K. Sarkar and O. Pereira, "Using the matrix pencil method to estimate the parameters of a sum of complex exponentials," *IEEE Antennas and Propagation Magazine*, vol. 37, pp. 48–55, Feb 1995. DOI: 10.1109/74.370583 182

[6] A. R. Weily and A. S. Mohan, "Full two-port analysis of coaxial-probe-fed TE01 and HE11 mode dielectric resonator filters using FDTD," *Microwave and Optical Technology Letters*, vol. 18, pp. 149–154, Jun 1998.
DOI: 10.1002/(SICI)1098-2760(19980605)18:2%3C149::AID-MOP18%3E3.0.CO;2-2 178

[7] K. Kurokawa, "Power waves and scattering matrix," *IEEE Transactions on Microwave Theory and Techniques*, vol. MT13, pp. 194-&, 1965. DOI: 10.1109/TMTT.1965.1125964 178

[8] C. A. Balanis, *Advanced Engineering Electromagnetics*. New York: Wiley, 1989. 186, 187, 188, 189

[9] J. A. Stratton, *Electromagnetic Theory*. New York: McGraw-Hill, 1941. 187

[10] D. J. Robinson and J. B. Schneider, "On the use of the geometric mean in FDTD near-to-far-field transformations," *IEEE Transactions on Antennas and Propagation*, vol. 55, pp. 3204–3211, Nov 2007. DOI: 10.1109/TAP.2007.908795 191

[11] A. F. Peterson, S. L. Ray, and R. Mittra, *Computational Methods for Electromagnetics*. New York: IEEE Press, 1998. 195

APPENDIX A

MATLAB Implementation of the 1D FDTD Model of a Uniform Transmission Line

A.1 TRANSLATING THE DISCRETE FDTD EQUATIONS TO A HIGH-LEVEL PROGRAMMING LANGUAGE

Chapter 2 studied with some detail the FDTD discretization of the one-dimensional transmission-line equations. The fundamental update equations for a lossless transmission line were presented in (2.13) and (2.14). Second-order accurate boundary conditions for a resistive load is provided by (2.59). A Thévenin voltage source excitation was given by (2.62). The discrete FDTD approximation will provide a second-order accurate time-dependent solution for the transmission line voltages and currents, providing the time-step satisfies the Courant Fredrichs-Lewy (CFL) stability limit (2.54). The topic of this Appendix is to develop an understanding of how to implement the discrete FDTD equations into an actual computer program. The language that has been chosen for this exercise is the MATLAB® programming language. This language has been chosen since it is widely used by university students. For those who do not have a license for MATLAB, the same code can be used with GNU Octave©. For flexibility and performance, a high-level language such as C++ or FORTRAN would be preferable. The codes provided herein can easily be translated to these languages.

To preserve second-order accuracy, the discrete voltages and currents are staggered by one-half of a cell in space and time. In this way, a discrete current sample is centered between two voltage samples, and *vice versa*. However, to directly implement the recursive update Equations (2.13) and (2.14) in a programming language, these equations must be re-written with purely integer indices. Consider the new indexing structure illustrated in Fig. A.1. The first and last voltage nodes are sampled at the source and load boundaries, respectively. As illustrated, the currents, centered between the voltage nodes, are sampled entirely within the interior of the transmission line. Let nx be the number of voltage nodes. Then, the discrete line voltages and currents are stored in two separate floating point arrays V and I of dimension nx and $nx-1$, respectively.

Observing Fig. A.1, the discrete update equations in (2.13) and (2.14) can be posed using discrete array indexing as:

$$V(k) = V(k) - \Delta t * (I(k) - I(k-1))/(C * \Delta x), \quad k = 2 : nx - 1 \qquad (A.1)$$

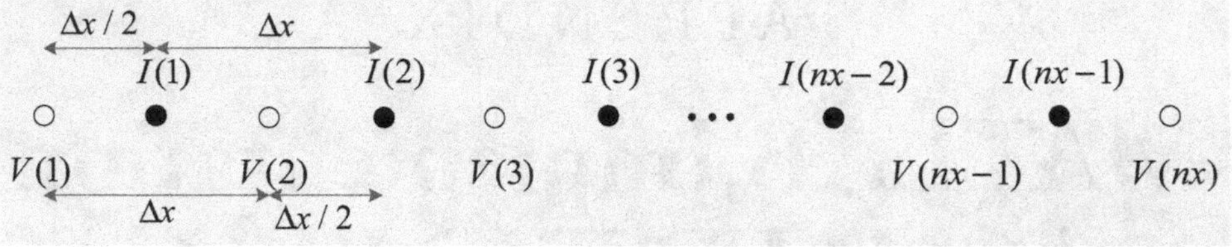

Figure A.1: Discrete array indexing of the line voltages and currents.

and

$$I(k) = I(k) - \Delta t * (V(k+1) - V(k))/(L * \Delta x), \quad k = 1 : nx - 1. \tag{A.2}$$

Note that the voltage update excludes the boundary nodes and is only valid over the range $k = 2 : nx - 1$. At the boundary nodes, separate update expressions must be used. For example, if the source is a Thevenin voltage source, then from (2.62), the discrete update can be expressed as:

$$V(1) = c1 * (c2 * V(1) - I(1) + V_g(n - 0.5 * \Delta t)/R_g)) \tag{A.3}$$

where

$$c1 = 1.0/(C * 0.5 * \Delta x/\Delta t + 0.5/R_g), \tag{A.4}$$
$$c2 = (C * 0.5 * \Delta x/\Delta t - 0.5/R_g) \tag{A.5}$$

and V_g is the known voltage source and is a function of time. Similarly, if the line is terminated by a resistive load, the load voltage update from (2.59) is expressed in discrete form as:

$$V(nx) = c1 * (c2 * V(nx) + I(nx - 1)) \tag{A.6}$$

where R_g is replaced with R_L in $c1$ and $c2$.

It is observed that there is no time dependence in the expressions in (A.1)–(A.6). The reason for this is that the expressions are implemented in a purely recursive manner. Given the initial conditions of the line voltages and currents, they are simply advanced recursively in time, overwriting their previous value at each discrete time update.

Figures A.2 through A.6 contain a simple set of routines written in MATLAB that compute the time-dependent voltages and currents on a uniform lossless transmission line that is excited by a Thévenin voltage source and terminated by a resistive load. Figure A.2 contains the main function. The user calling this function provides the per unit length parameters, the source and load values, as well as the number of cells and the CFL number. This function initializes the voltage and current arrays to zero for time $t \leq 0$. During every time-step, the interior line voltages are first updated. The boundary voltages are then updated separately. The currents are then updated at the next $1/2$ time-step. The time dependent voltages and currents at a list of points (vpts, ipts) are output to an external data file at each time-step. This solution scheme continues for all time-steps, at which time the simulation completes.

Figures A.3 and A.4 contain MATLAB functions that update the interior line voltages and currents, respectively, using (A.1) and (A.2). The functions updating the voltages at the source and load boundaries based on (A.3) and (A.6) are presented in Figures A.5 and A.6, respectively.

This program was used to generate the results in Figs. 2.7 and 2.8 the function Vg(t) was a unit amplitude trapezoidal pulse.

```
function txlineFDTD(C,L,d,ncells,cfln,simtime,Vg,Rg,RL,vpts,ipts)
% This function computes the FDTD analysis
% of a uniform transmission line
% excited by a Thevenin source and terminated by a resistive load.
%--------------------------------------------------------------- ---
% Input:
% C,L = per unit length capacitance & inductance (F/m, H/m)
% d = the length of the line (in m)
% ncells = the number of discrete cells defining the line
% cfnl = the CFL number
% simtime = the total simulation time (in s)
% Vg =reference to the function defining the voltage source Vg(t)
% Rg, RL = the source and load resistances, respectively,
% vpts, ipts = list of line voltage and current sample points for
% output
%-- ----------------------------- ------------------------------
    % compute the line parameters:
    Zo = sqrt(L/C); vo = 1.0/sqrt(L*C);

    % compute the discretization
    dx = d / ncells;
    dt = cfln * dx / vo;
    nx = ncells + 1; nt = floor(simtime / dt);

    % initialize the line voltages and currents to zero:
    V = zeros(nx,1); I = zeros(nx-1,1);
```

Figure A.2: Main MATLAB routine for the 1D FDTD simulation of the line voltages and currents on a uniform transmission line excited by a Thévenin source and terminated by a resistive load. (*Continues.*)

```
    % Time-advance the line voltages and currents with
    % in-line functions:
    for n = 1:nt
        % update the line voltages:
        vupdate;      % update interior node voltages
        vsupdate;     % update source node voltage
        vlupdate;     % update load node voltage

        % update the line currents:
        iupdate;

        % output the probe voltages and currents:
        output;
    end
end
```

Figure A.2: (*Continued.*) Main MATLAB routine for the 1D FDTD simulation of the line voltages and currents on a uniform transmission line excited by a Thévenin source and terminated by a resistive load.

```
% In-line function to compute the update of the line voltage
% interior to the homogeneous line
%-----------------------------------------------------------------
% compute the multiplier coefficient:

cv = dt/(C * dx);

% recursively update the line voltage at all interior nodes
% via (2.13):

for k=2:nx-1
    V(k) = V(k) - cv * (I(k) - I(k-1));
end
```

Figure A.3: MATLAB function that updates the interior voltages using (2.13).

```
% In-line function to compute the update of the line current
% interior to the homogeneous line
%-----------------------------------------------------------------
% compute the multiplier coefficient:

ci = dt/(L * dx);

% recursively update the line voltage at all interior nodes
% via (2.14):

for k=1:nx-1
    I(k) = I(k) - ci * (V(k+1) - V(k));
end
```

Figure A.4: MATLAB function that updates the interior current using (2.14).

```
% In-line function to update the source voltage V(1)
%----------------------------------------------------------------
% External Function:
% Vg = the source function
% n = the time-step
%----------------------------------------------------------------

% compute the open circuit source voltage:

Vs = Vg(n,dt);

if(Rg > 0)
% compute the multiplier coefficients (from (2.62)):
    b1 = C*dx*0.5/dt;
    b2 = 0.5/Rg;
    c1 = 1.0/(b1 + b2);
    c2 = b1 - b2;
    % update the line voltage at the source node
    V(1) = c1 * (c2 * V(1) - I(1) + Vs/Rg);
else
    V(1) = Vs;        % if Rs = 0, then V(1) = Vs
end
```

Figure A.5: MATLAB function that updates the source voltage using (2.62).

```
% In-line function to update the load voltage V(nx)
%---------------------------------------------- ----------------------
% Input:
% RL = the load resistance (should be >= 0).
%--------------------------------------------------------------------

% If the load resistance is 0, V(nx) = 0:
if(RL == 0)
    V(nx) = 0.;
else
% compute the multiplier coefficients (from (2.59)):
    b1 = C*dx*0.5/dt;
    b2 = 0.5/RL;
    c1 = 1.0/(b1 + b2);
    c2 = b1 - b2;
    % update the line voltage at the load:
    V(nx) = c1 * (c2 * V(nx) + I(nx-1));
end
```

Figure A.6: MATLAB function that updates the load voltage using (2.59).

APPENDIX B

Efficient Implementation of the 3D FDTD Algorithm

B.1   TOP-LEVEL DESIGN

The objective of this Appendix is to present efficient implementations of the FDTD algorithm using a high-level programming language. A proper program design should maintain the generality of the FDTD method, accommodating absorbing boundaries, general materials, conformal boundaries, subcell models, etc., without losing the efficiency of the base FDTD algorithm. These issues will be addressed throughout this Appendix.

Any good engineering design must start with a strong top-level design. The same is true with computer program design. This section focuses on the first step of designing the FDTD code – the top level design.

Figure B.1 illustrates a sample MATLAB code representative of the main function governing a full three-dimensional FDTD simulation. The code embodies the basic top-level design of the FDTD software. The first block of code reads in the problem dependent data sets, and then continues on to pre-process the mesh and then to pre-compute all the update coefficients needed for the FDTD simulation. The field arrays (and auxiliary variable arrays) are allocated and are initialized to zero as an initial condition.

The pre-processing phase is important, since pre-computing as much as possible *a priori* ultimately saves computational time during the more costly time-stepping phase. For even a modest number of time-steps, the time-update procedure dominates the overall computational time. Thus, an efficient implementation of the time-stepping algorithm is imperative.

At each time-step, the fields are recursively updated at all grid points throughout the domain. As will be seen in the following sections, the three-dimensional field updates can be efficiently expressed in a triple nested loop. The nested loops can be structured for optimal computational throughput. They are also highly parallelizable and can also be adapted to modern GPU-based computers as well.

An important element necessary for efficiency is that the same update equations should be used to update the fields throughout the entire 3D space. That is, the update loops should not be broken up with branch statements to select different update operators. Neither should different update loops be used to update fields in different regions. Rather, a single footprint should be used to update all fields throughout the entire volume. What will change in different regions will be the update coefficients. Based on this, the role of the functions `MainEFieldUpdate` and `MainHFieldUpdate` is to update

```
% Pre-processing:
%------------------------------------------------
% Input mesh dimensions, materials, conformal boundaries, sources,
% time-signature, simulation control, output control, etc.
ReadInputData;

% Initializations:
%------------------------------------------------
% pre-processing for subcell models such as conformal boundaries:
PreProcessSubCellGeometries;
% initialize the time-simulation (time-steps, sources, etc.)
InitilizeTimeSimulation;
% initialize PML parameters:
InitializePML;
% initialize all update coefficients (main & auxiliary, including PML):
InitializeUpdateCoefficients;
% intialize the fields and auxiliary variables throughout the grid:
InitializeFields;

% TimeStepping:
%------------------------------------------------
% Advance the fields through nMax time iterations:
for n = 1:nMax
    % Explicit update of the electric fields:
    %----------------------------------------
    % update electric fields throughout the entire domain:
    MainEFieldUpdate;
    % Auxiliary electric field updates (material, local, PML):
    AuxiliaryEFieldUpdates;
    % Outer boundary update of E-fields (ABC, or if PEC do nothing)
    OuterBoundaryEFieldUpdates;
    % Electric field source injection:
    ETypeSourceInjection;
```

Figure B.1: MATLAB code for the main FDTD program. (*Continues.*)

```
% Explicit update of the magnetic fields:
%----------------------------------------
% update magnetic fields throughout the entire domain:
MainHFieldUpdate;
% Auxiliary magnetic field updates (material, local, PML):
AuxiliaryHFieldUpdates;
% Magnetic field source injection:
HTypeSourceInjection;

% Field calculators:
%----------------------------------------------
% time-dependent field calculators (e.g., V, I, NF-FF)
FieldCalculators;

% Output Data
OutputData;
end
% Post processing:
%----------------------------------------------
% perform all post-processing calculations
PostProcess;
```

Figure B.1: (*Continued.*) MATLAB code for the main FDTD program.

the electric and magnetic fields, respectively, in all space. What differentiates the different model types will be the update coefficients. For example, the update equations of the fields in a free-space region (Section 3.2) will differ only from those in a lossy region (Section 3.8) by the coefficients weighting the previous field value and the discrete curl. Conformal FDTD formulations can also be written in a manner such that only the update coefficients need to be modified (c.f., (7.57) and (7.58)). Within a PEC volume, or on a PEC surface, the coefficients used to update the electric field should be set to zero. This explicitly enforces the zeroing of the electric field within the PEC volume, or tangential to the PEC surface for all time without disrupting the field updates. This will also force the magnetic fields within the PEC volume and normal to the PEC surface to be zero.

When auxiliary variables are introduced, such as in a PML region, or near a thin-sheet subcell model, a separate routine is called to update the auxiliary variables and to inject the results into the major fields. This is carried out by the routines AuxiliaryEFieldUpdates and

`AuxiliaryHFieldUpdates`. For example, in a PML region the updates of the auxiliary variables (c.f., (6.82)– (6.97)) will be performed in these routines. The auxiliary variables will also be injected back into the major fields. This would also be true for the field updates in a dispersive media. One of the advantages of using a stretched coordinate PML formulation is that the auxiliary field updates in the PML and the dispersive media regions are done completely independently. Even if they overlap, it does not impact the footprint of the top-level design.

At every time-step, the sources are injected into the fields. This is performed outside of the main field update loop via `ETypeSourceInjection` for electric sources and `HTypeSourceInjection` for magnetic sources. The sources are simply superimposed into the major fields within this routines.

The FieldCalculators block calculates indirect quantities, such as voltages, currents or near-field to far-field transforms, that must be updated at every time-step. The OutputData block dumps out user specified data to an external file (e.g., discrete fields, voltages or currents) at every time-step. The final block in the top-level design is the PostProcess block, which performs the necessary post processing operations (e.g., network parameter computation or far-fields).

The rest of the Appendix focuses on developing high-level programs for the main field updates.

B.2   ARRAY INDEXING THE 3D-FDTD

The FDTD discretization of the three-dimensional Maxwell's equations is presented in Chapter 3. The discrete update equations are derived via central difference approximations of both the time and space derivatives of Maxwell's curl equations. The resulting explicit update operators for the three-dimensional electric and magnetic fields for a lossless, homogeneous medium are presented in Equations (3.25) through (3.30). In order to translate these equations to a high-level programming language, the non-integer indexed fields in (3.25) through (3.30) must be mapped to integer-based array indices. To this end, consider the array indexing of the primary Yee-cell illustrated in Figs. B.2 and B.3. Figure B.2 presents the electric fields with integer indices projected on the edges of a primary Yee-cell with root node (i, j, k). Similarly, Fig. B.3 presents the magnetic fields with integer indices projected on the faces of the primary Yee-cell with root node (i, j, k). From (3.25), and Figs. B.2 and B.3, the update equation for H_x is written in array index form as:

$$H_x(i, j, k) = H_x(i, j, k)+$$
$$\frac{\Delta t}{\mu} \left[\frac{1}{\Delta z} \left(E_y(i, j, k + 1) - E_y(i, j, k) \right) - \frac{1}{\Delta y} \left(E_z(i, j + 1, k) - E_z(i, j, k) \right) \right]. \quad \text{(B.1)}$$

Assume that there are nx, ny, and nz nodes defining the global primary grid along the x, y, and z-directions, respectively. Then this update is valid for all the magnetic fields within the problem domain, which would include the range of indices $i = 1 : nx$, $j = 1 : ny - 1$, and $k = 1 : nz - 1$.

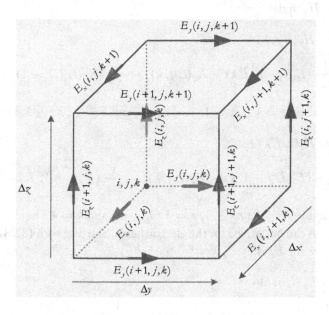

Figure B.2: Discrete array indexing of the electric fields of the unit primary Yee-cell.

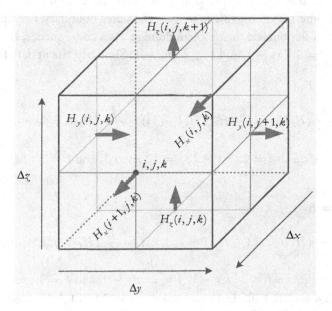

Figure B.3: Discrete array indexing of the magnetic fields of the unit primary Yee-cell.

Similarly, for the H_y update:

$$H_y(i, j, k) = H_y(i, j, k)+$$

$$\frac{\Delta t}{\mu} \left[\frac{1}{\Delta x} (E_z(i+1, j, k) - E_z(i, j, k)) - \frac{1}{\Delta z} (E_x(i, j, k+1) - E_x(i, j, k)) \right]. \quad \text{(B.2)}$$

for the range of indices $i = 1 : nx - 1$, $j = 1 : ny$, and $k = 1 : nz - 1$. For the H_z-update:

$$H_z(i, j, k) = H_z(i, j, k)+$$

$$\frac{\Delta t}{\mu} \left[\frac{1}{\Delta y} (E_x(i, j+1, k) - E_x(i, j, k)) - \frac{1}{\Delta x} \left(E_y(i+1, j, k) - E_y(i, j, k) \right) \right] \quad \text{(B.3)}$$

for the range of indices $i = 1 : nx - 1$, $j = 1 : ny - 1$, and $k = 1 : nz$.

Dual updates can be derived for the electric fields. Starting with (3.28) and observing Figs. B.2 and B.3, the update for E_x is:

$$E_x(i, j, k) = E_x(i, j, k)+$$

$$\frac{\Delta t}{\varepsilon} \left[\frac{1}{\Delta y} (H_z(i, j, k) - H_z(i, j-1, k)) - \frac{1}{\Delta z} \left(H_y(i, j, k) - H_y(i, j, k-1) \right) \right]. \quad \text{(B.4)}$$

It is observed that this update expression is not valid on the outer boundary (e.g., $j = 1$ or ny, or $k = 1$ or nz). The reason for this is that on the outer boundary, the update relies on a discrete magnetic field that lies outside the problem domain. As a consequence, the range of valid indices is $i = 1 : nx - 1$, $j = 2 : ny - 1$, and $k = 2 : nz - 1$. Similarly, the update for E_y is expressed as:

$$E_y(i, j, k) = E_y(i, j, k)+$$

$$\frac{\Delta t}{\varepsilon} \left[\frac{1}{\Delta z} (H_x(i, j, k) - H_x(i, j, k-1)) - \frac{1}{\Delta x} (H_z(i, j, k) - H_z(i-1, j, k)) \right] \quad \text{(B.5)}$$

for the range of indices is $i = 2 : nx - 1$, $j = 1 : ny - 1$, and $k = 2 : nz - 1$. The update for E_z is expressed as:

$$E_z(i, j, k) = E_z(i, j, k)+$$

$$\frac{\Delta t}{\varepsilon} \left[\frac{1}{\Delta x} \left(H_y(i, j, k) - H_y(i-1, j, k) \right) - \frac{1}{\Delta y} (H_x(i, j, k) - H_x(i, j-1, k)) \right] \quad \text{(B.6)}$$

for the range of indices is $i = 2 : nx - 1$, $j = 2 : ny - 1$, and $k = 1 : nz - 1$.

Figures B.4 and B.5 illustrate codes written in MATLAB® used to update the magnetic and electric fields based on (B.1)–(B.3) and (B.4)–(B.6), respectively. (MATLAB syntax is chosen due to its popularity and simplicity. The codes can readily be translated into other programming languages such as FORTRAN, C, or C++.) The definitions of the constant coefficients can be derived

```
% Main update loops for the magnetic fields:
for k = 1:nz-1
    for j = 1:ny-1
        for i = 1:nx
            Hx(i,j,k) = Hx(i,j,k)-cHxy*(Ez(i,j+1,k)-Ez(i,j,k))+
                                  cHxz*(Ey(i,j,k+1)-Ey(i,j,k));
        end
    end
end

for k = 1:nz-1
    for j = 1:ny
        for i = 1:nx-1
            Hy(i,j,k) = Hy(i,j,k)-cHyz*(Ex(i,j,k+1)-Ex(i,j,k))+
                                  cHyx*(Ez(i+1,j,k)-Ez(i,j,k));
        end
    end
end

for k = 1:nz
    for j = 1:ny-1
        for i = 1:nx-1
            Hz(i,j,k) = Hz(i,j,k)-cHzx*(Ey(i+1,j,k)-Ey(i,j,k))+
                                  cHzy*(Ex(i,j+1,k)-Ex(i,j,k));
        end
    end
end
```

Figure B.4: MATLAB code for updating the magnetic fields throughout the solution domain for a lossless homogeneous medium and a uniformly spaced grid based on (B.1)–(B.3).

from (B.1)–(B.6). For example, in Fig. B.4 $cHxy = \Delta t/(\mu \Delta y)$, the coefficients are assumed to be pre-computed by `InitializeUpdateCoefficients` and stored in memory.

The updates of the tangential electric fields on the outer domain boundary are performed separately. If the outer boundary is a perfect electric conductor, no update is performed. Rather, the tangential fields on the outer boundary are initialized to 0, and, consequently, they remain zero

```
% Main update loops for the electric fields:
for k = 2:nz-1
    for j = 2:ny-1
        for i = 1:nx-1
            Ex(i,j,k) = Ex(i,j,k)+cExy*(Hz(i,j,k)-Hz(i,j-1,k))-
                                  cExz*(Hy(i,j,k)-Hy(i,j,k-1));
        end
    end
end

for k = 2:nz-1
    for j = 1:ny-1
        for i = 2:nx-1
            Ey(i,j,k) = Ey(i,j,k)+cEyz*(Hx(i,j,k)-Hx(i,j,k-1))-
                                  cEyx*(Hz(i,j,k)-Hz(i-1,j,k));
        end
    end
end

for k = 1:nz-1
    for j = 2:ny-1
        for i = 2:nx-1
            Ez(i,j,k) = Ez(i,j,k)+cEzx*(Hy(i,j,k)-Hy(i-1,j,k))-
                                  cEzy*(Hx(i,j,k)-Hx(i,j-1,k));
        end
    end
end
```

Figure B.5: MATLAB code for updating the electric fields throughout the solution domain for a lossless homogeneous medium and a uniformly spaced grid based on (B.4)–(B.6).

through the computation. If an absorbing boundary condition (ABC) is used to terminate the outer boundary, then the appropriate boundary condition (c.f., Chapter 5) is used to update the tangential electric fields on the outer boundary.

MATLAB uses column-wise storage of multi-dimensional arrays (similar to FORTRAN). Thus, the triple-nested loops in Figs. B.3 and B.4 are structured to access the field arrays sequentially in memory. The innermost loops cycle over index i – which is the inner index of the field arrays. The two external loops cycle over the outer indexes j and k, respectively. Most modern processors rely heavily on cache to maintain a proper throughput. During the execution, the processor pre-fetches contiguous blocks of memory of the arrays being operated on and pulls them into cache. While executing the updates on the blocks of arrays in cache, the next block of memory is pre-fetched. Thus, by organizing the loops in this way, the array elements needed for the updates will more likely be in cache, optimizing the floating-point performance.

B.3 LOSSY AND INHOMOGENEOUS MEDIA

The program presented in the previous section can be generalized further to the case of a lossy, inhomogeneous media. The update equation for lossy media was presented in (3.99) for the electric field. Using Figs. B.2 and B.3, this can be rewritten using array indexes as:

$$E_x(i, j, k) = cex1(i, j, k)E_x(i, j, k) + cex2(i, j, k) \cdot$$

$$[(Hz(i, j, k) - Hz(i, j - 1, k))/\Delta y - (Hy(i, j, k) + Hy(i, j, k - 1))/\Delta z] \quad \text{(B.7)}$$

where the coefficients are computed as:

$$cex1(i, j, k) = \frac{2\varepsilon_x(i, j, k) - \sigma_x(i, j, k)\Delta t}{2\varepsilon_x(i, j, k) + \sigma_x(i, j, k)\Delta t}, \quad \text{(B.8)}$$

$$cex2(i, j, k) = \frac{2\Delta t}{2\varepsilon_x(i, j, k) + \sigma_x(i, j, k)\Delta t} \quad \text{(B.9)}$$

and $\varepsilon_x(i, j, k)$ is the averaged dielectric constant and $\sigma_x(i, j, k)$ is the averaged conductivity about the edge associated with $E_x(i, j, k)$ (c.f., Section 3.7). The coefficients are pre-computed and stored. Update expressions for E_y and E_z are similarly posed. Note, in general, $\sigma_x(i, j, k)$ and $\varepsilon_x(i, j, k)$ need not be explicitly stored as three-dimensional arrays. Rather, their values can be computed "on-the-fly" while pre-computing the arrays $cex1$ and $cex2$.

Pre-computing and storing the coefficients, $cex1$ and $cex2$, saves a significant amount of CPU time. Since they are three-dimensional arrays, the cost of storage is quite expensive since six arrays will be needed for the electric field updates. This can be alleviated if the material conforms to a separable geometry in a Cartesian coordinate system. For example, if the medium is a layered medium, then the arrays can be reduced to one-dimensional arrays. For more general geometries, three-dimensional indexing is required. However, in general, the number of distinct values of $cex1$ and $cex2$ is much smaller than the number of grid edges. Thus, the coefficients $cex1$ and $cex2$ can be stored in reduced dimension one-dimensional arrays that are indirectly addressed. In this way, (B.7) can be written as:

$$E_x(i, j, k) = cexa(iex(i, j, k))E_x(i, j, k) + cexb(iex(i, j, k)) \cdot$$

$$[(Hz(i, j, k) - Hz(i, j - 1, k))/\Delta y - (Hy(i, j, k) + Hy(i, j, k - 1))/\Delta z] \quad \text{(B.10)}$$

where iex is an integer array, and $cexa$ and $cexb$ are one-dimensional floating point arrays. The array iex is set up such that $cexa(iex(i, j, k)) = cex1(i, j, k)$ and $cexb(iex(i, j, k)) = cex2(i, j, k)$. The dimension of $cexa$ and $cexb$ is equal to the number of distinct values of $cex1$ and $cex2$. In general, these arrays will be small. Thus, the cost of storing the coefficients is reduced by a factor of two.

Similar arrays can be developed for the E_y and E_z updates. It is further noted that a single integer array can be use to address all six coefficient arrays. This can increase the size of the one-dimensional coefficient arrays. However, in general, this increase is going to be much smaller than the reduction of the memory from storing six three-dimensional coefficient arrays, to six one-dimensional arrays of dramatically reduced dimension, plus a single three-dimensional integer array.

B.4   IMPLEMENTING THE CFS-CPML

The CPML is the most commonly used form of the PML. This is mostly due to its accuracy and efficiency, as well as its material independence. The theory of the CFS-CPML was outlined in Chapter 6. The update equations derived for the CFS-PML are given by (6.75)–(6.97). For example, consider the update of E_x in (6.78) for a lossless medium. Using Figs. B.2 and B.3, this equation is expressed in array notation as:

$$E_x(i, j, k) = E_x(i, j, k) + \frac{\Delta t}{\varepsilon \kappa_y \Delta y} (H_z(i, j, k) - H_z(i, j - 1, k)) -$$

$$\frac{\Delta t}{\varepsilon \kappa_z \Delta z} (H_y(i, j, k) - H_y(i, j, k - 1)) + \frac{\Delta t}{\varepsilon} \left(Q_{y,z}^H(i, j, k) - Q_{z,y}^H(i, j, k) \right). \quad \text{(B.11)}$$

Similarly, the updates for the auxiliary variables are given by (6.92) and (6.93). These are expressed using array index notation as:

$$Q_{y,z}^H(i, j, k) = b_y(j) Q_{y,z}^H(i, j, k) - \frac{c_y(j)}{\Delta y} (H_z(i, j, k) - H_z(i, j - 1, k)) \quad \text{(B.12)}$$

and

$$Q_{z,y}^H(i, j, k) = b_z(k) Q_{z,y}^H(i, j, k) - \frac{c_z(k)}{\Delta z} (H_y(i, j, k) - H_z(i, j, k - 1)). \quad \text{(B.13)}$$

The arrays $Q_{y,z}^H(i, j, k)$ and $Q_{z,y}^H(i, j, k)$ only need to be stored in the PML region. Outside the PML region, $Q_{y,z}^H(i, j, k)$ and $Q_{z,y}^H(i, j, k)$ are zero, $\kappa_y = \kappa_z = 1$, and (B.11) reduces to the standard update (B.4).

To implement the CFS-PML within an FDTD code, one could apply (B.11) in the PML region, and (B.4) outside the PML using separate loops. However, splitting the updates in this manner would be inefficient. A better alternative is to first update E_x in all space using a variant of (B.11) that closely resembles (B.4):

$$E_x(i, j, k) = E_x(i, j, k) + \frac{\Delta t}{\varepsilon \kappa_y \Delta y} \left(H_z(i, j, k) - H_z(i, j-1, k)\right) -$$

$$\frac{\Delta t}{\varepsilon \kappa_z \Delta z} \left(H_y(i, j, k) - H_y(i, j, k-1)\right) \tag{B.14}$$

and superimpose the auxiliary terms in a separate loop in `AuxiliaryEFieldUpdates` via:

$$E_x(i, j, k) = E_x(i, j, k) + \frac{\Delta t}{\varepsilon} \left(Q_{y,z}^H(i, j, k) - Q_{z,y}^H(i, j, k)\right). \tag{B.15}$$

Figure B.6 presents MATLAB code to perform the electric field updates (B.14) throughout the entire volume (inside and outside of the PML regions). The update coefficients are computed from (6.79)–(6.81). For example, `cexz(j)=`$\Delta t/(\kappa_y(j)\Delta y\varepsilon)$ is computed at the edge center. The remaining coefficients are similarly computed. Note that the electric field updates of all three electric fields are updated in a single triple-nested loop, rather than three loops, as done in Fig. B.6. This reduces the number of loop operations. It also improves the floating point throughput by increasing redundancy of data in cache. The tradeoff of putting all the fields in a single loop structure is that the full range of each update is not spanned. The remaining field blocks are updated in separate double-nested loops (i.e., for sub blocks i =1 (for E_x), j =1 (for E_y) and k =1 (for E_z)). The cost of the supplemental two-dimensional loops is small compared to the triple-nested loop, and thus is an acceptable tradeoff.

The MATLAB code used to update the auxiliary fields in `AuxiliaryEFieldUpdates` and superimpose these fields into the electric field in the PML region is illustrated in Fig. B.7. Six different loops are explicitly written out for each of the six boundary PML layers, since each layer can have separate thicknesses. For example, `npmlx_1` defines the number of cells across the left most PML (starting with i = 1). Similarly, `npmlx_2` is the number of cells through the PML layer along the x-direction for the right-most boundary (ending with $i = nx$).

The first loop structure in Fig. B.7 loops over the left x-PML layer. In this loop, both $Q_{x,z}^H$ and $Q_{x,y}^H$ are updated. Both auxiliary variables share identical update coefficients b_x and c_x. Within the loop, $Q_{x,z}^H$ is superimposed into E_y, and $Q_{x,y}^H$ is superimposed into E_z. From (6.79) and (6.80) `coef_e` = $\Delta t/\varepsilon$ for a homogeneous lossless space. It is also observed that $Q_{x,z}^H$ and $Q_{x,y}^H$ are not updated at the outer boundary (i= 1) since a PEC boundary is assumed to terminate the outer boundary, and the tangential electric fields and Q's are zero on this boundary.

The second loop structure in Fig. B.7 loops over the right x-PML layer. Again, both $Q_{x,z}^H$ and $Q_{x,y}^H$ are updated in this loop. Note that these are stored in compressed arrays that only have dimension `npmlx_2` along the x-direction. Both $Q_{x,z}^H$ and $Q_{x,y}^H$ are first updated, and then superimposed into the electric fields E_y and E_z, respectively. Finally, $Q_{x,z}^H$ and $Q_{x,y}^H$ are not updated on the outer boundary (i = nx) since the tangential electric fields and $Q's$ are zero on this boundary.

```
% Main update loop for the E-field (Lossless, homogeneous, media)
% throughout all space, including the PML region.
for k = 2:nz-1
    for j = 2:ny-1
        for i = 2:nx-1
            Ex(i,j,k) = Ex(i,j,k)+(Hz(i,j,k)-Hz(i,j-1,k))*cexz(j)-
                                    (Hy(i,j,k)-Hy(i,j,k-1))*cexy(k);
            Ey(i,j,k) = Ey(i,j,k)+(Hx(i,j,k)-Hx(i,j,k-1))*ceyx(k)-
                                    (Hz(i,j,k)-Hz(i-1,j,k))*ceyz(i);
            Ez(i,j,k) = Ez(i,j,k)+(Hy(i,j,k)-Hy(i-1,j,k))*cezy(i)-
                                    (Hx(i,j,k)-Hx(i,j-1,k))*cezx(j);
        end
    end
end
% Remaining E-field updates near the exterior boundary
i = 1;
for k = 2:nz-1
    for j = 2:ny-1
        Ex(i,j,k) = Ex(i,j,k)+(Hz(i,j,k)-Hz(i,j-1,k))*cexz(j)-
                                (Hy(i,j,k)-Hy(i,j,k-1))*cexy(k);
    end
end
j = 1;
for k = 2:nz-1
    for i = 2:nx-1
        Ey(i,j,k) = Ey(i,j,k)+(Hx(i,j,k)-Hx(i,j,k-1))*ceyx(k)-
                                (Hz(i,j,k)-Hz(i-1,j,k))*ceyz(i);
    end
end
k = 1;
for j = 2:ny-1
    for i = 2:nx-1
        Ez(i,j,k) = Ez(i,j,k)+(Hy(i,j,k)-Hy(i-1,j,k))*cezy(i)-
                                (Hx(i,j,k)-Hx(i,j-1,k))*cezx(j);
    end
end
```

Figure B.6: MATLAB code in `MainEFieldUpdate` for updating the electric fields throughout the solution with PML scaling as well as for non-uniform grid.

```
% Update loops for the auxiliary variables and superimposing the
% updated auxiliary variables into the electric field:
%--------------------
% left x-pml region:
for k = 2:nz-1
    for j = 2:ny-1
        for i = 2,npmlx_1+1
            Q_Hxz_1(i,j,k) = bx_e_1(i)*Q_Hxz_1(i,j,k)+
                                    cx_e_1(i)*(Hz(i,j,k)-Hz(i-1,j,k));
            Q_Hxy_1(i,j,k) = bx_e_1(i)*Q_Hxy_1(i,j,k)+
                                    bx_e_1(i)*(Hy(i,j,k)-Hy(i-1,j,k));
            Ey(i,j,k) = Ey(i,j,k)-Q_Hxz_1(i,j,k)*coef_e;
            Ez(i,j,k) = Ez(i,j,k)+Q_Hxy_1(i,j,k)*coef_e;
        end
    end
end

% right x-pml region
for k = 2:nz-1
    for j = 2:ny-1
        i = nx - npmlx_2
        for i2 = 1,npmlx_2
            Q_Hxz_2(i2,j,k) = bx_e_2(i2)*Q_Hxz_2(i2,j,k)+
                                    cx_e_2(i2)*(Hz(i,j,k)-Hz(i-1,j,k));
            Q_Hxy_2(i2,j,k) = bx_e_2(i2)*Q_Hxy_2(i2,j,k)+
                                    bx_e_2(i2)*(hy(i,j,k)-Hy(i-1,j,k));
            Ey(i,j,k) = Ey(i,j,k)-Q_Hxz_2(i2,j,k)*coef_e;
            Ez(i,j,k) = Ez(i,j,k)+Q_Hxy_2(i2,j,k)*coef_e;
            i = i + 1;
        end
    end
end
```

Figure B.7: MATLAB code within `AuxiliaryEFieldUpdates` for updating the auxiliary variables and adding the superimposed terms to the electric field within the x, y, and z-normal PML regions. (*Continues.*)

```
%---------------------
% left y-pml region:
for k = 2:nz-1
    for j = 2:npmly_1+1
        for i = 2,nx-1
            Q_Hyx_1(i,j,k) = by_e_1(i)*Q_Hyx_1(i,j,k)+
                            cy_e_1(i)*(Hx(i,j,k)-Hx(i,j-1,k));
            Q_Hyz_1(i,j,k) = by_e_1(i)*Q_Hyz_1(i,j,k)+
                            by_e_1(i)*(Hz(i,j,k)-Hz(i,j-1,k));
            Ez(i,j,k) = Ez(i,j,k)-Q_Hyx_1(i,j,k)*coef_e;
            Ex(i,j,k) = Ex(i,j,k)+Q_Hyz_1(i,j,k)*coef_e;
        end
    end
end

% right xypml region
for k = 2:nz-1
    j = ny - npmly_2
    for j2 = 1: npmly_2
        for i = 2,nx-1
            Q_Hyx_2(i,j2,k) = by_e_2(j2)*Q_Hyx_2(i,j2,k)+
                             cy_e_2(j2)*(Hx(i,j,k)-Hx(i,j-1,k));
            Q_Hyz_2(i,j2,k) = by_e_2(j2)*Q_Hyz_2(i,j2,k)+
                             by_e_2(j2)*(Hz(i,j,k)-Hz(i,j-1,k));
            Ez(i,j,k) = Ez(i,j,k)-Q_Hyx_2(i,j2,k)*coef_e;
            Ex(i,j,k) = Ex(i,j,k)+Q_Hyz_2(i,j2,k)*coef_e;
            j = j + 1;
        end
    end
end
```

Figure B.7: (*Continued.*) MATLAB code within `AuxiliaryEFieldUpdates` for updating the auxiliary variables and adding the superimposed terms to the electric field within the x, y, and z-normal PML regions. (*Continues.*)

```
%--------------------
% left z-pml region:
for k = 2:npmlz_1+1
    for j = 2:ny-1
        for i = 2,nx-1
            Q_Hzy_1(i,j,k) = bz_e_1(i)*Q_Hzy_1(i,j,k)+
                             cz_e_1(i)*(Hy(i,j,k)-Hy(i,j,k-1));
            Q_Hzx_1(i,j,k) = bz_e_1(i)*Q_Hzx_1(i,j,k)+
                             bz_e_1(i)*(Hx(i,j,k)-Hx(i,j,k-1));
            Ex(i,j,k) = Ex(i,j,k)-Q_Hzy_1(i,j,k)*coef_e;
            Ey(i,j,k) = Ey(i,j,k)+Q_Hzx_1(i,j,k)*coef_e;
        end
    end
end

% right z-pml region
k = nz - npmlz_2
for k = 1:npmlz_2
    for j2 = 2:ny-1
        for i = 2,nx-1
            Q_Hzy_2(i,j,k2) = bz_e_2(k2)*Q_Hzy_2(i,j,k2)+
                              cz_e_2(k2)*(Hy(i,j,k)-Hy(i,j,k-1));
            Q_Hzx_2(i,j,k2) = bz_e_2(k2)*Q_Hzx_2(i,j,k2)+
                              bz_e_2(k2)*(Hx(i,j,k)-Hx(i,j,k-1));
            Ex(i,j,k) = Ex(i,j,k)-Q_Hzy_2(i,j,k2)*coef_e;
            Ey(i,j,k) = Ey(i,j,k)+Q_Hzx_2(i,j,k2)*coef_e;
            k = k + 1;
        end
    end
end
```

Figure B.7: (*Continued.*) MATLAB code within `AuxiliaryEFieldUpdates` for updating the auxiliary variables and adding the superimposed terms to the electric field within the x, y, and z-normal PML regions.

```
% Main update loop for the H-field (Lossless, homogeneous, media)
% throughout all space, including the PML region.
for k = 1:nz-1
    for j = 1:ny-1
        for i = 1:nx-1
            Hx(i,j,k) = Hx(i,j,k)-(Ez(i,j+1,k)-Ez(i,j,k))*chxz(j)+
                                   (Ey(i,j,k+1)-Ey(i,j,k))*chxy(k);
            Hy(i,j,k) = Hy(i,j,k)-(Ex(i,j,k+1)-Ex(i,j,k))*chyx(k)+
                                   (Ez(i+1,j,k)-Ez(i,j,k))*chyz(i);
            Hz(i,j,k) = Hz(i,j,k)-(Ey(i+1,j,k)-Ey(i,j,k))*chzy(i)+
                                   (Ex(i,j+1,k)-Ex(i,j,k))*chzx(j);
        end
    end
end
```

Figure B.8: MATLAB code in `MainHFieldUpdate` for updating the magnetic fields throughout the solution with PML scaling as well as for non-uniform grid.

The remaining loops update the auxiliary variables in the left and right y- and z-normal PML regions. In each loop, two auxiliary variables are updated, and then superimposed into the appropriate electric field.

The magnetic fields are similarly updated via superposition. First, the implementation of the global magnetic field updates is illustrated in Fig. B.8. Again, this is expressed in a single three-dimensional loop structure. Note that the supplemental two-dimensional loops needed to update the magnetic fields on the extreme outer boundary are not needed. The reason for this is that the outer boundary is assumed to be a PEC, and the normal magnetic field is zero on these boundaries.

The update of the auxiliary variables and the superposition of these variables into the magnetic field is illustrated in Fig. B.9. Only the left and right x-normal PML regions are illustrated. The y- and z-normal PML regions can be similarly derived.

Next, consider the computation of the update coefficients. Computing these coefficients is somewhat challenging because the PML constitutive parameters are spatially scaled along the normal axis of the PML. It is for this reason that the coefficients are expressed as one-dimensional arrays. The appropriate means for computing the coefficients is to compute them at the edge or face center of the electric or magnetic field that they are updating.

Figures B.10 and B.11 illustrate MATLAB functions that compute the update coefficients for the electric and magnetic field updates, respectively. Note that each function can be used to compute

```
% Update loops for the auxiliary variables and superimposing the
% updated auxiliary variables into the magnetic field:
% left x-pml region:
for k = 2:nz-1
    for j = 2:ny-1
        for i = 1,npmlx_1
            Q_Exz_1(i,j,k) = bx_h_1(i)*Q_Exz_1(i,j,k)+
                             cx_h_1(i)*(Ez(i+1,j,k)-Ez(i,j,k));
            Q_Exy_1(i,j,k) = bx_h_1(i)*Q_Exy_1(i,j,k)+
                             bx_h_1(i)*(Ey(i+1,j,k)-Ey(i-1,j,k));
            Hy(i,j,k) = Hy(i,j,k)-Q_Exz_1(i,j,k)*coef_h;
            Hz(i,j,k) = Hz(i,j,k)+Q_Exy_1(i,j,k)*coef_h;
        end
    end
end

% right x-pml region
for k = 2:nz-1
    for j = 2:ny-1
        i = nx - npmlx_2
        for i2 = 1,npmlx_2
            Q_Exz_2(i2,j,k) = bx_h_2(i2)*Q_Exz_2(i2,j,k)+
                              cx_h_2(i2)*(Ez(i+1,j,k)-Ez(i,j,k));
            Q_Exy_2(i2,j,k) = bx_h_2(i2)*Q_Exy_2(i2,j,k)+
                              bx_h_2(i2)*(Ey(i+1,j,k)-Ey(i,j,k));
            Hy(i,j,k) = Hy(i,j,k)-Q_Exz_2(i2,j,k)*coef_h;
            Hz(i,j,k) = Hz(i,j,k)+Q_Exy_2(i2,j,k)*coef_h;
            i = i + 1;
        end
    end
end
```

Figure B.9: MATLAB code within `AuxiliaryHFieldUpdates` for updating the auxiliary variables and adding the superimposed terms to the electric field within the x-normal PML region only.

```
function [kappa bcoef ccoef] =
    compE_PMLCoefs(iend,del,dt,eps,npml,m,ma,sigmax,kapmax,alpmax)

% Function used to compute the PML coefficient arrays used for
% e-field updates in the PML region.
%------------------------------------------------------------
% Input:
% iend = 1 (left side), or = 2 (right side)
% del = the grid cell spacing along this direction (dx,dy, or dz)
% dt, eps = delta_t and eps of the host media (typically eps0)
% npml = the number of pml cells
% m = polynomial scaling for sigma and kappa
% ma = polynomial scaling for alpha
% sigmax,kapmax,alpmax = max values for sigma, kappa, and alpha
%------------------------------------------------------------

% left-side coefficient computation:
if(iend == 1) then
    for i = 1:npml+1
        polyscalesk = ((npml+1.0-i)/npml)^m;
        polyscalea = ((i -1.0)/npml)^ma;
        sigma(i) = sigmax * polyscalesk;
        kappa(i) = 1.0 + (kapmax-1.0) * polyscalesk;
        alpha(i) = alpmax * polyscalea;
    end
else
    for i = 1:npml+1
        polyscalesk = ((i -1.0)/npml)^m;
        polyscalea = ((npml+1.0-i)/npml)^ma;
        sigma(i) = sigmax* polyscalesk;
        kappa(i) = 1.0 + (kapmax-1.0)* polyscalesk;
        alpha(i) = alpmax * polyscalea;
    end
end
```

Figure B.10: MATLAB coding for scaled PML tensor coefficients and the computing the update coefficients within the PML region for e-updates. (*Continues.*)

```
for i = 1:npml+1
    temp = kappa(i)*alpha(i)+sigma(i);
    dtotau = temp*dt/(kappa(i)*eps);
    bcoef(i) = exp(-dtotau);

    if(temp == 0.0)
        ccoef(i) = 0.0;
    else
        ccoef(i) = sigma(i)*(bcoef(i)-1.0)/(kappa(i)*temp*del);
    end
end
```

Figure B.10: (*Continued.*) MATLAB coding for computing the scaled PML tensor coefficients and the update coefficients within the PML region for e-updates.

the coefficients for any of the six PML boundary layers. For example, the coefficients for the left x-PML layer can be computed via the function in Fig. B.10 using the command:

$$
\begin{aligned}
& \texttt{[kappa\_ex\_1 bx\_e\_1 cx\_e\_1] =} \\
& \quad \texttt{compE\_PMLCoefs(1,dx,dt,eps0,npmlx\_1,m,ma,sigmax, kapmax,alpmax)}
\end{aligned}
\tag{B.16}
$$

where the first argument (`iend`) is set to 1 to indicate the left-boundary. The coefficients `bx_e_1` and `cx_e_1` arc used for the field updates in the first loop structure in Fig. B.7. Also, `kappa_ex_1` is used to compute the coefficients `ceyz(i)` and `cezy(i)` in Fig. B.6. For the electric field updates, the coefficients are used to update the fields that are orthogonal to the PML axis. Consequently, the constitutive parameters and the update coefficients are computed at the nodes.

The MATAB function used to compute the magnetic field update coefficients is illustrated in Fig. B.11. For example, the coefficients for the left x-PML layer can be computed via the command:

$$
\begin{aligned}
& \texttt{[kappa\_hx\_1 bx\_h\_1 cx\_h\_1] =} \\
& \quad \texttt{compH\_PMLCoefs(1,dx,dt,eps0,npmlx\_1,m,ma, sigmax,kapmax,alpmax)}
\end{aligned}
\tag{B.17}
$$

The coefficients `bx_h_1` and `cx_h_1` are used for the field updates in the first loop structure in Fig. B.9. Also, `kappa_hx_1` is used to compute the coefficients `chyz(i)` and `chzy(i)`. These coefficients are used to update the magnetic fields that are orthogonal to the PML axis. Consequently, the constitutive parameters and the update coefficients, are computed at half-cell increments, as observed in Fig. B.11.

```matlab
function [kappa bcoef ccoef] =
    compH_PMLCoefs(iend,del,dt,eps,npml,m,ma,sigmax,kapmax,alpmax)

% Function used to compute the PML coefficient arrays used for
% e-field updates in the PML region.
%-------------------------------------------------------------
% Input:
% iend = 1 (left side), or = 2 (right side)
% del = the grid cell spacing along this direction (dx,dy, or dz)
% dt, eps = delta_t and eps of the host media (typically eps0)
% npml = the number of nodes in the pml region (including bdries)
% m = polynomial scaling for sigma and kappa
% ma = polynomial scaling for alpha
% sigmax,kapmax,alpmax = max values for sigma, kappa, and alpha
%-------------------------------------------------------------

% left-side coefficient computation:
if(iend == 1) then
    for i = 1:npml
        polyscalesk = ((npml-i+0.5)/npml)^m;
        polyscalea = ((i - 0.5)/npml)^ma;
        sigma(i) = sigmax * polyscalesk;
        kappa(i) = 1.0 + (kapmax-1.0) * polyscalesk;
        alpha(i) = alpmax * polyscalea;
    end
else
    for i = 1:npml
        polyscalesk = ((i - 0.5)/npml)^m;
        polyscalea = ((npml-i+0.5)/npml)^ma;
        sigma(i) = sigmax * polyscalesk;
        kappa(i) = 1.0 + (kapmax-1.0)* polyscalesk;
        alpha(i) = alpmax * polyscalea;
    end
end
```

Figure B.11: MATLAB coding for computing the scaled PML tensor coefficients and the update coefficients within the PML region for e-updates. (*Continues.*)

```
for i = 1:npml-1
    temp = kappa(i)*alpha(i)+sigma(i);
    dtotau = temp*dt/(kappa(i)*eps);
    bcoef(i) = exp(-dtotau);

    if(temp == 0.0)
        ccoef(i) = 0.0;
    else
        ccoef(i) = sigma(i)*(bcoef(i)-1.0)/(kappa(i)*temp*del);
    end
end
```

Figure B.11: (*Continued.*) MATLAB coding for computing the scaled PML tensor coefficients and the update coefficients within the PML region for e-updates.

B.5   EDGE LENGTH NORMALIZATION

In a typical FDTD simulation, 99% of the CPU time is spent computing the updates in (B.1)–(B.6). Consequently, the updates must be implemented as efficiently as possible to minimize CPU time. One way this has been done so far was to put all three field updates within a single triple-nested loop and to properly structure the loops. Another way was to reduce the total number of floating point operations by scaling the fields by their respective edge lengths. To this end, let

$$e_x(i, j, k) = E_x(i, j, k)\Delta x, \quad e_y(i, j, k) = E_y(i, j, k)\Delta y, \quad e_z(i, j, k) = E_z(i, j, k)\Delta z \quad (B.18)$$

and

$$h_x(i, j, k) = H_x(i, j, k)\Delta x, \quad h_y(i, j, k) = H_y(i, j, k)\Delta y, \quad h_z(i, j, k) = H_z(i, j, k)\Delta z. \quad (B.19)$$

With this scaling the units of the scaled electric and magnetic fields are volts and amperes, respectively.
    Thus, from (B.1), the update for the scaled magnetic field h_x is:

$$h_x(i, j, k) = h_x(i, j, k) + \frac{\Delta t}{\mu} \frac{\Delta x}{\Delta y \Delta z} \left[e_y(i, j, k + 1) - e_y(i, j, k) - e_z(i, j + 1, k) + e_z(i, j, k) \right].$$
$$(B.20)$$

Similar expressions are derived for h_y and h_z. From (B.4),

$$e_x(i, j, k) = e_x(i, j, k) + \frac{\Delta t}{\varepsilon} \frac{\Delta x}{\Delta y \Delta z} \left[h_z(i, j, k) - h_z(i, j - 1, k) - h_y(i, j, k) + h_y(i, j, k - 1) \right].$$
$$(B.21)$$

It is interesting to note that the resulting expressions are also consistent with those derived via the finite integration technique presented in Section 3.4. The coefficients can be pre-computed and stored. Thus, comparing (B.20) and (B.21) with (B.1) and (B.4), respectively, the total number of floating point operations is reduced by 17%.

B.6   GENERAL FDTD UPDATE EQUATIONS

As discussed in Section B.1, the objective is to provide a single footprint that can efficiently update the electric and magnetic fields throughout the entire grid. This would include fields in inhomogeneous or dispersive media regions, the PML region, non-uniform grid regions, on conformal boundaries, or fields impacted by subcell models. In fact, all the field update equations derived for a wide variety of FDTD methods were presented in a manner that are "similar to the Yee-update equations." The rationale for this was to provide such a footprint. In fact, this foot print is provided by (B.20) and (B.21) and can be expressed in the form of:

$$e_x(i, j, k) = e_x(i, j, k) + cexb(i, j, k) \cdot \left[h_z(i, j, k) - h_z(i, j-1, k) - h_y(i, j, k) + h_y(i, j, k-1) \right]$$
(B.22)

and

$$h_x(i, j, k) = h_x(i, j, k) + chxb(i, j, k) \cdot \left[e_y(i, j, k+1) - e_y(i, j, k) - e_z(i, j+1, k) + e_z(i, j, k) \right]$$
(B.23)

for lossless media, and

$$e_x(i, j, k) = cexa(i, j, k) \cdot e_x(i, j, k) + cexb(i, j, k)$$
$$\cdot \left[h_z(i, j, k) - h_z(i, j-1, k) - h_y(i, j, k) + h_y(i, j, k-1) \right]$$
(B.24)

and

$$h_x(i, j, k) = chxa(i, j, k) \cdot h_x(i, j, k) + chxb(i, j, k)$$
$$\cdot \left[e_y(i, j, k+1) - e_y(i, j, k) - e_z(i, j+1, k) + e_z(i, j, k) \right]$$
(B.25)

for lossy media. How this is done will be explored for a few sets of field update equations. The remaining will be left for an exercise. This leads to a purely general implementation of the FDTD method using a single footprint.

Initially, we will start with a non-uniform gridding as discussed in Section 3.10. Applying the normalizations:

$$e_x(i, j, k) = E_x(i, j, k)\Delta x(i), \ e_y(i, j, k) = E_y(i, j, k)\Delta y(j), \ e_z(i, j, k) = E_z(i, j, k)\Delta z(k)$$
(B.26)

and

$$h_x(i, j, k) = H_x(i, j, k)\ell_x(i), \ h_y(i, j, k) = H_y(i, j, k)\ell_y(j), \ h_z(i, j, k) = H_z(i, j, k)\ell_z(k),$$
(B.27)

then from (3.116) and (3.115), this leads to the lossless media updates (B.22) and (B.23), where

$$cexb(i, j, k) = \frac{\Delta t}{\varepsilon(i, j, k)} \cdot \frac{\Delta x(i)}{\ell_y(j)\ell_z(k)} \qquad \text{(B.28)}$$

and

$$chxb(i, j, k) = \frac{\Delta t}{\mu(i, j, k)} \cdot \frac{\ell_x(i)}{\Delta y(j)\Delta z(k)}. \qquad \text{(B.29)}$$

Next, consider the PML media. For simplicity, we will assume that the host medium is lossless. In the PML medium, we apply the normalizations

$$e_x(i, j, k) = E_x(i, j, k)\kappa_x(i)\Delta x, \quad e_y(i, j, k) = E_y(i, j, k)\kappa_y(j)\Delta y,$$
$$e_z(i, j, k) = E_z(i, j, k)\kappa_z(k)\Delta z \qquad \text{(B.30)}$$

and, similarly

$$h_x(i, j, k) = H_x(i, j, k)\kappa_x(i)\Delta x, \quad h_y(i, j, k) = H_y(i, j, k)\kappa_y(j)\Delta y,$$
$$h_z(i, j, k) = H_z(i, j, k)\kappa_z(k)\Delta z. \qquad \text{(B.31)}$$

Then, from (6.78), the main electric field update is expressed via (B.22) with

$$cexb(i, j, k) = \frac{\Delta t}{\varepsilon(i, j, k)} \cdot \frac{\kappa_x(i)\Delta x}{\kappa_y(j)\Delta y\kappa_z(k)\Delta z} \qquad \text{(B.32)}$$

and the updated auxiliary variable is superimposed into the electric field separately as:

$$e_x(i, j, k) = e_x(i, j, k) + \frac{\Delta t\kappa_x(i)\Delta x}{\varepsilon(i, j, k)}\left(Q_{y,z}^H(i, j, k) - Q_{z,y}^H(i, j, k)\right). \qquad \text{(B.33)}$$

The auxiliary variables are updated in terms of the normalized fields as:

$$Q_{y,z}^H(i, j, k) = b_y(j)Q_{y,z}^H(i, j, k) - \frac{c_y(j)}{\Delta y\Delta z}\frac{1}{\kappa_z(k)}(h_z(i, j, k) - h_z(i, j-1, k)) \qquad \text{(B.34)}$$

and

$$Q_{z,y}^H(i, j, k) = b_z(k)Q_{z,y}^H(i, j, k) - \frac{c_z(k)}{\Delta y\Delta z}\frac{1}{\kappa_y(j)}(H_y(i, j, k) - H_z(i, j, k-1)). \qquad \text{(B.35)}$$

Similar updates are derived for the magnetic field.

Now consider the Dey-Mittra (DM) conformal grid FDTD method for a closed PEC object, as detailed in Section 7.3.1. The updates equations for the electric and magnetic fields are provided by (7.58) and (7.57), respectively. Note that electric fields have already been normalized as detailed by (7.55). After applying the scaling (B.18) to the normalized electric fields (7.55) and (B.19) to the magnetic fields, this leads to update equations identical to (B.22) and (B.23), with coefficients given by (B.28)

$$cexb(i, j, k) = \frac{\Delta t}{\varepsilon(i, j, k)} \cdot \frac{\Delta x}{\Delta y \Delta z} \tag{B.36}$$

and

$$chxb(i, j, k) = \frac{\Delta t}{\mu(i, j, k)} \cdot \frac{\Delta x}{\Delta y \Delta z} \tag{B.37}$$

where $\varepsilon(i, j, k)$ and $\mu(i, j, k)$ are given by (7.59) and by (7.56). Thus, the DM conformal FDTD algorithm can be expressed in a form such that the main field update algorithms conform to (B.22) and (B.23).

Author's Biography

STEPHEN D. GEDNEY

Stephen D. Gedney is a Professor of Electrical and Computer Engineering at the University of Kentucky, Lexington, KY, where he has been since 1991. He was named the Reese Terry Professor of Electrical and Computer Engineering at the University of Kentucky in 2002. He received the B.Eng.-Honors degree from McGill University, Montreal, Q.C., in 1985, and the M.S. and Ph.D. degrees in Electrical Engineering from the University of Illinois, Urbana-Champaign, IL, in 1987 and 1991, respectively. He has been a NASA/ASEE Faculty Fellow with the Jet Propulsion Laboratory, Pasadena, CA. He has also served as a visiting research engineer with the Hughes Research Labs (now HRL laboratories) in Malibu, CA, and Alpha Omega Electromagnetics, Ellicott City, MD. He is also the recipient of the Tau Beta Pi Outstanding Teacher Award. He is a Fellow of the IEEE.

Prof. Gedney's research is in the area of computational electromagnetics with focus on the finite-difference time-domain, discontinuous Galerkin time-domain methods, high-order solution algorithms, fast solver technology, and parallel algorithms. His research has focused on applications in the areas of electromagnetic scattering and microwave circuit modeling and design. He has published over 150 articles in peer reviewed journals and conference proceedings and has contributed to a number of books in the field of computational electromagnetics.

Index

Printed in the United States
by Baker & Taylor Publisher Services